AF595469

Nutritional and Antioxidant Value of Horticulturae Products

Nutritional and Antioxidant Value of Horticulturae Products

Editors

Lucia Guidi
Luigi De Bellis
Alberto Pardossi

MDPI • Basel • Beijing • Wuhan • Barcelona • Belgrade • Manchester • Tokyo • Cluj • Tianjin

Editors
Lucia Guidi
University of Pisa
Italy

Luigi De Bellis
Università del Salento
Italy

Alberto Pardossi
University of Pisa
Italy

Editorial Office
MDPI
St. Alban-Anlage 66
4052 Basel, Switzerland

This is a reprint of articles from the Special Issue published online in the open access journal *Horticulturae* (ISSN 2311-7524) (available at: https://www.mdpi.com/journal/horticulturae/special_issues/Nutritional_Antioxidant_Value).

For citation purposes, cite each article independently as indicated on the article page online and as indicated below:

LastName, A.A.; LastName, B.B.; LastName, C.C. Article Title. *Journal Name* **Year**, *Volume Number*, Page Range.

ISBN 978-3-0365-3785-6 (Hbk)
ISBN 978-3-0365-3786-3 (PDF)

Cover image courtesy of Lucia Guidi

Contents

About the Editors

Lucia Guidi, Professor of Agricultural Biochemistry, degree in Agricultural Science with top score, PhD in Protein and Oil Crops, University of Pisa: Director of the Interdepartmental Research Center Nutrafood "Nutraceuticals and Food for Health", at the University of Pisa from 2018 until 2024. Author of over 170 scientific publications with a *h* index of 41. Main scientific interests: photosynthesis (gas exchange and chlorophyll fluorescence) in plants under stress; secondary metabolism (metabolites and pathway); plant response abiotic and biotic stress.

Luigi De Bellis, Professor of Plant Physiology, degree in Agricultural Science with top score, Ph.D. in Crop and Fruit Trees Science (curriculum Propagation), University of Pisa. Post Doc in Japan (twice) in the National Institute for Basic Biology (NIBB), Okazaki, in the lab directed by Prof. M. Nishimura, and at The University of Edinburgh, Scotland, UK, in the Dr. S.M. Smith's lab. Chair Professor of Plant Physiology at the University of Salento, Lecce, Italy from November 1st, 2002. Member of the Editorial board of Biology and Plants; from January 2022 Editor in Chief of Horticulturae. Author of over 100 scientific publications. Main scientific interests: enzymes of the glyoxylate cycle; intracellular localization of aconitase isoforms; carbohydrates and control of gene expression; genetics of plant species such as olive trees and wheat; *Xylella fastidiosa* as olive tree pest and selection of tolerant olive varieties; heavy metal uptake and translocation in plants; plant phenolic compounds and volatiles.

Alberto Pardossi, Professor of Vegetable and Ornamental Crops at the University of Milan (1988–2001) and the University of Pisa (2001–); degree in Agricultural Science with tope score; PhD in Horticulture, University of Pisa; Postdoc at the School Biological Sciences, University of Bangor, NW, UK (1991); Coordinator of the PhD Program in Agriculture, Food and Environment at the University of Pisa (2009–2016); Director of the Department of Agriculture, Food and Environment at the University of Pisa (2016–2020). Member of the Editorial Board of Horticulturae, Advances in Horticultural Sciences, and Italian Journal of Agronomy. Author of 149 indexed (SCOPUS) scientific publications with a *h* index of 34. Main scientific interests: greenhouse horticulture; hydroponics; plant water relations and mineral nutrition; plant response to abiotic and biotic stress.

MDPI

Editorial

Nutritional and Antioxidant Value of Horticulturae Products

Lucia Guidi [1,2,*], Luigi De Bellis [3,*] and Alberto Pardossi [1,2,*]

1 Department of Agriculture, Food and Environment, University of Pisa, I-56124 Pisa, Italy
2 Interdepartmental Research Center Nutrafood "Nutraceuticals and Food for Health", University of Pisa, I-56124 Pisa, Italy
3 Department of Biological and Environmental Sciences and Technologies, University of Salento, I-73100 Lecce, Italy
* Correspondence: lucia.guidi@unipi.it (L.G.); luigi.debellis@unisalento.it (L.D.B.); alberto.pardossi@unipi.it (A.P.)

The recent growing interest towards the nutraceutical and antioxidant value of fruit and vegetables has arisen from their content of phytochemicals, which provide desirable health benefits, beyond basic nutrition, to reduce the risk of major chronic diseases [1]. For this reason, it is necessary for the human diet to contain a good proportion of plant antioxidant compounds. Therefore, horticultural science must dedicate more attention to satisfy the expectations of consumers who are demanding more and more high-quality functional foods.

The Special Issue on the "Nutritional and Antioxidant Value of Horticultural Products" has provided readers with novel insights into some 'unusual' types of foods. In fact, the articles published included information about edible flowers, Italian green tea, and stinging nettle, along with potato and sweet potato.

Demasi et al. [2] investigated the sensory profile of 17 edible flowers (Figure 1) at harvest, and their shelf life and bioactive compound dynamics during cold storage, providing the characteristics and requirements of the different flowers. The authors evaluated the aroma of the flower with 10 sensory descriptors (intensity of sweet, sour, bitter, salt, smell, specific flower aroma, and herbaceous aroma; spiciness, chewiness, and astringency), and, both at harvest and during 14 d of storage at 4 °C, the flower visual quality, polyphenol and anthocyanin content, and antioxidant activity were evaluated. The paper presents a lot of information concerning flowers' sensory profiles, phytochemical characteristics, and shelf life, which are very useful to select suitable species for the edible flower market; for example, a strong aroma was revealed for *Allium ursinum*, *Dianthus carthusianorum*, *Lavandula angustifolia*, and *Leucanthemum vulgare*, while the flowers with the longest shelf life were *Rosa pendulina* (14 days) and *Rosa canina* (10 days).

Instead, at the University of Pisa (Italy) [3], different post-harvest treatments (air drying at 30, 50, 60, and 70 °C, and freeze drying) were applied to *Agastache aurantiaca* (A.Gray) Lint & Epling flowers to determine the effects on the bioactive metabolites and volatile compounds. The outcome was that freeze-drying was found to be the best solution to prolong the shelf life and keep the antioxidant compounds of *A. aurantiaca* flowers, despite changes in the aromatic component. In fact, pulegone, the main volatile emitted by fresh flowers, decreases after the different drying treatments while other volatile compounds appeared.

Drying was also applied to stinging nettle (*Urtica dioica* L.) [4], which is a well-known plant in traditional medicine. Because the technique enables a low-cost product to be stored for long time, dehydrated samples, after freeze-drying (FD), oven-drying (OD), and heat pump drying (HPD), and fresh leaves were subjected to water extraction to emulate their use by the final consumers. HPD gave the best results, even doubling the phenol content and antioxidant activity compared to the fresh sample, but with a reduction of around 40% in the ascorbic acid content and 10 to over a hundred times higher amounts of some specific phenolic compounds. These findings confirm the value of dehydration techniques, which, however, should always be modulated when used on fresh samples.

Citation: Guidi, L.; De Bellis, L.; Pardossi, A. Nutritional and Antioxidant Value of Horticulturae Products. *Horticulturae* **2022**, *8*, 4. https://doi.org/10.3390/horticulturae8010004

Received: 17 November 2021
Accepted: 24 November 2021
Published: 21 December 2021

Figure 1. Images of seventeen edible flowers selected by Demasi et al. [2]. From left to right, first line is as follows: *Allium ursinum* L., *Borago officinalis* L., *Calendula officinalis* L., *Centaurea cyanus* L., *Cichorium intybus* L., *Dianthus carthusianorum* L.; second line is as follows: Lavandula angustifolia Mill., *Leucanthemum vulgare* (Vaill.) Lam., *Paeonia officinalis* L., *Primula veris* L., *Robinia pseudoacacia* L., *Rosa canina* L.; third line is as follows: *Rosa pendulina* L., *Salvia pratensis* L., *Sambucus nigra* L., Taraxacum officinale Weber, *Tropaeolum majus* L.

In Italy, in the Lake Maggiore district, the cultivation of *Camelia sinensis* (L.) O. Kuntze has developed for the production of a quality tea for a niche market. Therefore, the bioactive compounds (polyphenols) and antioxidant activity of the two green teas (the "Camellia d'Oro" tea—TCO, and the "Compagnia del Lago" tea—TCL) were analyzed [5]. The antioxidant activity of the two teas was similar or higher than in other green teas, with TCO characterized by higher antioxidant activity and phenolic content than TCL, indicating that it is possible to grow *C. sinensis* in Italy to a tea of satisfactory quality.

Tilahun et al. [6] realized a study to develop indices to predict the level of total glycoalkaloids (GA) in potato tubers at different greening stages, because the toxicity and beneficial effects of glycoalkaloids depend on the dose and conditions of use. The indices were based on surface color and chlorophyll content for estimation of safe GA levels, which represents basic information that needs to be confirmed and thereafter adapted for use on all potato cultivars.

Finally, Galvao et al. [7] evaluated the quality and nutrient contents of sweet potato, which is considered, by the FAO, as a primary crop for "traditional agriculture", and is also characterized by high variability concerning the color and nutritional value of the edible root. The analyses of 29 sweet potato accessions confirmed that there is considerable variability in nutrient content related to color. Deep-orange-fleshed sweet potatoes contain a higher level of β-carotene compared to light-orange- and cream-fleshed ones; therefore, 100 g of edible product can supply 32.3% of the daily value of vitamin A. The total phenolic content of the purple ecotypes was also higher than the other genotypes. Such high variability in the quality and nutrient contents suggests novel uses of the different color accessions in Europe as well, where sweet potato consumption is not widespread.

Today's food products are all the more appreciated for their content of phyto-nutriceuticals, and, therefore, horticulture has to turn to the supply of products with a variety of valuable metabolites and aromas. It is the task of researchers to study these compounds and enable the enhancement of horticultural products.

Author Contributions: All authors contributed equally. All authors have read and agreed to the published version of the manuscript.

Funding: This research received no external funding

Institutional Review Board Statement: Not applicable.

Informed Consent Statement: Not applicable.

Conflicts of Interest: The authors declare no conflict of interest.

References

1. Liu, R.H. Potential synergy of phytochemicals in cancer prevention: Mechanism of action. *J. Nutr.* **2004**, *134*, 3479S–3485S. [CrossRef] [PubMed]
2. Demasi, S.; Mellano, M.G.; Falla, N.M.; Caser, M.; Scariot, V. Sensory Profile, Shelf Life, and Dynamics of Bioactive Compounds during Cold Storage of 17 Edible Flowers. *Horticulturae* **2021**, *7*, 166. [CrossRef]
3. Marchioni, I.; Dimita, R.; Gioè, G.; Pistelli, L.; Ruffoni, B.; Pistelli, L.; Najar, B. The Effects of Post-Harvest Treatments on the Quality of *Agastache aurantiaca* Edible Flowers. *Horticulturae* **2021**, *7*, 83. [CrossRef]
4. Garcìa, L.M.; Ceccanti, C.; Negro, C.; De Bellis, L.; Incrocci, L.; Pardossi, A.; Guidi, L. Effect of Drying Methods on Phenolic Compounds and Antioxidant Activity of *Urtica dioica* L. Leaves. *Horticulturae* **2021**, *7*, 10. [CrossRef]
5. Falla, N.M.; Demasi, S.; Caser, M.; Scariot, V. Phytochemical Profile and Antioxidant Properties of Italian Green Tea, a New High Quality Niche Product. *Horticulturae* **2021**, *7*, 91. [CrossRef]
6. Tilahun, S.; An, H.S.; Solomon, T.; Baek, M.W.; Choi, H.R.; Lee, H.C.; Jeong, C.S. Indices for the Assessment of Glycoalkaloids in Potato Tubers Based on Surface Color and Chlorophyll Content. *Horticulturae* **2020**, *6*, 107. [CrossRef]
7. Galvao, A.C.; Nicoletto, C.; Zanin, G.; Vargas, P.F.; Sambo, P. Nutraceutical Content and Daily Value Contribution of Sweet Potato Accessions for the European Market. *Horticulturae* **2021**, *7*, 23. [CrossRef]

 *horticulturae*

Article

Sensory Profile, Shelf Life, and Dynamics of Bioactive Compounds during Cold Storage of 17 Edible Flowers

Sonia Demasi, Maria Gabriella Mellano, Nicole Mélanie Falla, Matteo Caser and Valentina Scariot *

Department of Agricultural, Forest and Food Sciences, University of Turin, 10095 Turin, Italy; sonia.demasi@unito.it (S.D.); gabriella.mellano@unito.it (M.G.M.); nicolemelanie.falla@unito.it (N.M.F.); matteo.caser@unito.it (M.C.)
* Correspondence: valentina.scariot@unito.it; Tel.: +39-011-6708-932

Abstract: In this study, 17 edible flowers (*Allium ursinum* L., *Borago officinalis* L., *Calendula officinalis* L., *Centaurea cyanus* L., *Cichorium intybus* L., *Dianthus carthusianorum* L., *Lavandula angustifolia* Mill., *Leucanthemum vulgare* (Vaill.) Lam., *Paeonia officinalis* L., *Primula veris* L., *Robinia pseudoacacia* L., *Rosa canina* L., *Rosa pendulina* L., *Salvia pratensis* L., *Sambucus nigra* L., *Taraxacum officinale* Weber, and *Tropaeolum majus* L.) were investigated to assess their sensory profile at harvest and their shelf life and bioactive compounds dynamics during cold storage. The emerging market of edible flowers lacks this information; thus, the characteristics and requirements of different flower species were provided. In detail, a quantitative descriptive analysis was performed by trained panelists at flower harvest, evaluating 10 sensory descriptors (intensity of sweet, sour, bitter, salt, smell, specific flower aroma, and herbaceous aroma; spiciness, chewiness, and astringency). Flower visual quality, biologically active compounds content (total polyphenols and anthocyanins), and antioxidant activity (FRAP, DPPH, and ABTS assays) were evaluated both at harvest and during storage at 4 °C for 14 days to assess their shelf life. Generally, species had a wide range of peculiar sensory and phytochemical characteristics at harvest, as well as shelf life and bioactive compounds dynamics during postharvest. A strong aroma was indicated for *A. ursinum*, *D. carthusianorum*, *L. angustifolia*, and *L. vulgare*, while *B. officinalis* and *C. officinalis* had very low values for all aroma and taste descriptors, resulting in poor sensory profiles. At harvest, *P. officinalis*, *R. canina*, and *R. pendulina* exhibited the highest values of polyphenols (884–1271 mg of gallic acid equivalents per 100 g) and antioxidant activity (204–274 mmol Fe^{2+}/kg for FRAP, 132–232 and 43–58 μmol of Trolox equivalent per g for DPPH and ABTS). The species with the longest shelf life in terms of acceptable visual quality was *R. pendulina* (14 days), followed by *R. canina* (10 days). All the other species lasted seven days, except for *C. intybus* and *T. officinale* that did not reach day 3. During cold storage, the content of bioactive compounds differed, as total phenolics followed a different trend according to the species and anthocyanins remained almost unaltered for 14 days. Considering antioxidant activity, ABTS values were the least variable, varying in only four species (*A. ursinum*, *D. carthusianorum*, *L. angustifolia*, and *P. officinalis*), while both DPPH and FRAP values varied in eight species. Taken together, the knowledge of sensory profiles, phytochemical characteristics and shelf life can provide information to select suitable species for the emerging edible flower market.

Keywords: anthocyanins; aroma; flavor; polyphenols; sensory analysis; postharvest; shelf life

Citation: Demasi, S.; Mellano, M.G.; Falla, N.M.; Caser, M.; Scariot, V. Sensory Profile, Shelf Life, and Dynamics of Bioactive Compounds during Cold Storage of 17 Edible Flowers. *Horticulturae* **2021**, *7*, 166. https://doi.org/10.3390/horticulturae7070166

Academic Editors: Luigi De Bellis, Lucia Guidi and Alberto Pardossi

Received: 12 May 2021
Accepted: 25 June 2021
Published: 29 June 2021

1. Introduction

The consumption of flowers as food is an ancient practice but many flowers, or parts of them, have had a much wider use in the past than today [1–6]. Rose petals (*Rosa* spp.) were already used in Roman times as ingredients in various preparations, as well as chamomile (*Matricaria chamomilla* L.) in ancient Greece and chrysanthemum (*Chrysanthemum morifolium* Ramat.) in China. In the Middle Ages, common marigold flowers (*Calendula officinalis* L.) were used as components of salads, especially in France; in the same region from the 1600s onwards, various products based on violets (*Viola odorata* L.) became popular. Similarly, in various areas of Europe, carnation (*Dianthus caryophyllus* L.), dandelion (*Taraxacum officinale* Weber) and elder (*Sambucus nigra* L.) flowers were consumed. Some of these food cultures that were once confined to rural populations have survived, albeit marginally, to the present day and have recently been revived adding color, flavor, taste and visual appeal to food preparations [7,8].

Today, the assortment of edible flowers includes several species that are used to improve the aesthetic appearance, color, and aroma of foods but also for their nutritional properties [2,9–13]. Edible flowers contain indeed several bioactive compounds (vitamins, minerals, phenolic substances), while they are poor in fat and proteins [2,3,9,11,13–16]. Many studies foster the nutritional interest in wild and ornamental flowers, similar to leafy vegetables. Phenolic compounds (e.g., phenolic acids, flavonoids, and anthocyanins) are among the most representative biologically active compounds as they are a rich family of phytochemicals. Additionally, antioxidant effects [17–20] are highly correlated in edible flowers [1]. These compounds exert several biological activities important for human health [7,20,21]. Phenolic compounds counteract oxidative stress caused by reactive oxygen species [18], and epidemiological data showed that a diet rich in antioxidants could prevent chronic diseases, such as cancer, cardiovascular and neurodegenerative disorders [2,18,22–24].

The increasing demand for more attractive, tasty and healthy food can lead to the production of edible flowers to complement growers' revenues, creating opportunities to develop value-added products in the floriculture sector, facing a challenging market [1,21,23]. Nowadays, several flowers are available on the market [25], however being few comparing to the variety of the species with edible flowers. It is therefore important to widen the knowledge about their quality, phytochemical composition and marketability to face the demand of consumers, producers, and retailers.

The sensory properties of food are extremely important not just to consumers, but also to food producers, because they relate directly to product quality and end-user acceptance [26], particularly concerning unfamiliar food, such as edible flowers [8,27]. According to the ISO 9000:2015 on quality management systems, the quality is the degree to which a combination of characteristics fulfils requirements [26]. Concerning edible flowers, sensory attributes such as color, appearance, flavor, and texture should be included in these characteristics [28,29]. Sensory science is a scientific discipline that concerns the presentation of a stimulus to subjects and then the evaluation of the subjects' response [30]. Studies on sensory profiles or aptitudes of consumers towards edible flowers are increasing [8,27,31–35] but only a few were performed by trained panelists [29,36].

Edible flowers are highly perishable products, with early petal abscission and discoloration, flower wilt, dehydration, and tissue browning [11,37,38]. After harvesting, plant organs continue living, and both respiration and transpiration processes are considered the major causes of postharvest losses and poor quality [39]. Senescence is controlled by developmental [40,41] and environmental signals [42]. Among environmental factors, temperature plays a major role in slowing down these processes, affecting the metabolism of harvested flowers and their shelf life [11,38,43]. Temperatures from −2.5 °C to 20 °C differently affected the quality and appearance of edible flowers according to the species, showing the possibility to extend their shelf life by decreasing the temperature of storage [11]. Amid low temperatures, the values often chosen are in the range 4–6 °C [2,37,39,44–46] and the most frequently evaluated parameter during postharvest has been the visual quality so far. Thus, further detailed studies are needed to understand the dynamics of bioactive compounds in edible flowers and their antioxidant activity upon cold storage.

Recently, we characterized several species of fresh edible flowers by means of spectrometry and chromatography, discovering a wide range of variability among species and numerous promising sources of bioactive compound, such as roses (*Rosa canina* L. and *Rosa pendulina* L.), peony (*Paeonia officinalis* L.), or *Primula veris* L. [13]. This research provides the sensory profiles of the flowers of 17 species, performed by trained panelists to add information about this emerging type of food. Their shelf life during storage at 4 °C for 14 days was then assessed through visual quality evaluation, as well as the content of their biologically active compounds (total polyphenols and anthocyanins) and antioxidant activity (FRAP, DPPH, and ABTS assays) both at harvest and during cold storage to evaluate their quality and marketability.

2. Materials and Methods

2.1. Plant Material

Seventeen edible flower species (Figure 1) were selected (Table 1) for the sensory and postharvest evaluation, including different properties and uses [13] and a wide assortment of flower color, shape and aroma, i.e., the traits that mostly attract consumers to try edible flowers [8]. Flowers were collected in the nursery for the species already available on the market (e.g., *B. officinalis*, *C. officinalis*, *L. angustifolia*, and *T. majus*), while for the others, it was necessary to collect them from wild plants. Flowers were collected at their full bloom (March through September according to the species), in 2017 and 2018. See reference [13] for detailed information on sampling sites and month. Flowers were preserved in plastic boxes inside a portable refrigerator. A portion of the sample was subjected to sensory evaluation within a few hours and another portion was transported to the laboratories of the Department of Agricultural, Forest and Food Sciences (DISAFA—University of Torino; Long. 7.589, Lat. 45.066) for analyses and postharvest trial.

Figure 1. Seventeen edible flowers selected for the study. From left to right, first line: *Allium ursinum* L., *Borago officinalis* L., *Calendula officinalis* L., *Centaurea cyanus* L., *Cichorium intybus* L., *Dianthus carthusianorum* L.; second line: *Lavandula angustifolia* Mill., *Leucanthemum vulgare* (Vaill.) Lam., *Paeonia officinalis* L., *Primula veris* L., *Robinia pseudoacacia* L., *Rosa canina* L.; third line: *Rosa pendulina* L., *Salvia pratensis* L., *Sambucus nigra* L., *Taraxacum officinale* Weber, *Tropaeolum majus* L.

2.2. Sensory Analysis

2.2.1. Panel Member Selection and Training

The sensory analysis was performed by the Italian National Organization of Fruit Tasters (O.N.A.Frut), composed of highly trained panelists that have been working since 2001 to promote and valorize sensory evaluation in the fruit sector. Panelists were trained in sensory analysis and quantitative descriptive analysis, being able to differentiate between basic taste solutions and aromas at various levels. The training on edible flowers started in 2016 with twenty people to improve their perception sensitivity and evaluation of individual descriptors, according to ISO 8586:2012 and ISO 3972:2011.

Table 1. List of the 17 species of edible flowers studied in the present work, with related beneficial properties and food use reported in the literature.

Species (Common Name)	Flower Properties	Eaten in/as	References
Allium ursinum L. (wild garlic)	Antioxidant, anti-inflammatory, antimycotic, cardioprotective.	Garlic substitute.	[47–49]
Borago officinalis L. (borage)	Purifying, emollient, antitussive, diuretic, sudorific, anti-inflammatory	Salads, soups, desserts, syrups and drinks. Cucumber taste.	[2,15,50]
Calendula officinalis L. (calendula)	Anti-inflammatory, antispasmodic, antiseptic, hepatoprotective, emollient, refreshing, cicatrizing.	Flavoring and decoration of salted dishes, bakery products and herb teas. Food coloring.	[2,51,52]
Centaurea cyanus L. (cornflower)	Diuretic, anti-inflammatory, disinfectant.	Garnishing dishes, syrups, teas	[2]
Cichorium intybus L. (chicory)	Laxative, diuretic, hypoglycemic, depurative, disinfectant, hepatoprotective.	Salads, soups.	[18]
Dianthus carthusianorum L. (Carthusian pink)	Diuretic, sudorific, nervine stimulant, febrifuge, sedative.	Infusions, liquors.	[53]
Lavandula angustifolia Mill. (lavender)	Antispasmodic, antiseptic, sedative, carminative, cicatrizing.	Flavoring and decoration of cakes, soups, salads, jellies. Essential oil to flavor food.	[15,54]
Leucanthemum vulgare Lam. (ox-eye daisy)	Antispasmodic, diuretic, tonic, antifungal, antibacterial.	Tea, salads	[55,56]
Paeonia officinalis L. (common peony)	Antirheumatic, antispasmodic, anti-inflammatory, analgesic, hepatoprotective.	Infusions.	[57,58]
Primula veris L. (cowslip)	Anti-inflammatory, anti-viral, anti-asthmatic.	Garnishing dishes, conserves, salads	[55,59]
Robinia pseudoacacia L. (black locust)	Antispasmodic, antiviral, diuretic, emollient, febrifuge, laxative, purgative, tonic.	Flavoring liquors, jams, honey, pancakes.	[18,55]
Rosa canina L. (dog rose)	Anticancer, diuretic, laxative, anti-rheumatic, anti-inflammatory.	Salads, jellies, syrups, teas.	[2,60]
Rosa pendulina L. (Alpine rose)	Anticancer, diuretic, laxative, anti-rheumatic.	Salads, jellies.	[2]
Salvia pratensis L. (meadow sage)	Anti-inflammatory, antibacterial, antiseptic, eupeptic.	Flavoring of butter, vinegar, oil, salads and creams, soups. Essential oil to flavor food.	[61]
Sambucus nigra L. (elder)	Antioxidant, anti-inflammatory, antibacterial, diuretic, emollient, sudorific, laxative, cardioprotective.	Herb teas and drinks. Flavoring honey, jellies and jams. Salads.	[18]
Taraxacum officinale Weber (dandelion)	Antioxidant, anti-inflammatory, hepatoprotective, diuretic, laxative, depurative, analgesic.	Salads and soups.	[62,63]
Tropaeolum majus L. (nasturtium)	Disinfectant, antimicrobial, expectorant, diuretic, anti-inflammatory.	Salads, flavoring of soups, meat, pasta, cheese, vinegar. Peppery flavor.	[2,64,65]

To guarantee a common lexicon of organoleptic terminology, the judges worked for four weeks tasting flowers and evaluating the samples both in groups and individually during the training sessions. After each panel session, a discussion was held according to literature [28,66–69] to define the descriptors in terms of appearance, aroma, texture and taste, following bibliographic references, to build an evaluation sheet for the Quantitative Descriptive Analysis (QDA). Of the 20 participants, twelve subjects, including males and females aged 20 to 60, were selected to form the panel and analyze five species (*C. officinalis*, *L. vulgare*, *R. pseudoacacia*, *S. nigra*, *T. majus*) in 2017 and twelve species (*A. ursinum*, *B. officinalis*, *C. cyanus*, *C. intybus*, *D. carthusianorum*, *L. angustifolia*, *P. officinalis*, *P. veris*, *R. canina*, *R. pendulina*, *S. pratensis*, *T. officinale*) in 2018.

2.2.2. Sensory Evaluation Test

The sessions for the sensory evaluation of the edible flowers were carried out in the sensory laboratory of O.N.A.Frut (Cuneo Province, Italy). Each judge received about 10 g of species items that had been presented as follows: about 5 g of flowers in a glass for olfactory evaluation and about 5 g in a white plastic dish for visual and tasting evaluation. Each species was evaluated independently by each panelist. According to sensory analysis rules, the sample presentation was basic, without other food in order to uniform the total impact of different species. Flowers were evaluated in the harvest day, fresh and without cooking preparation. All samples were served in duplicate to all judges and the order of presentation was randomized within each test day. Between tasting, assessors were encouraged to clean their palates with water during a 5-min break.

2.2.3. Quantitative Descriptive Analysis (QDA)

The Quantitative Descriptive Analysis (QDA) is a key part of sensory methodology: only when the intensity of sensory traits is rated, a food can be described in detail regarding its taste. The QDA method joins descriptor intensity points together with visually display difference [70]. The QDA was applied as analytical-descriptive method for sensory evaluation of flower samples. Each selected descriptor was evaluated on a continuous scale partially structured into 10 segments with intervals from zero (absence of the character) to 10 (maximum intensity). The evaluations of this study were based on previous experiences of taste evaluation performed on vegetables and fruits [69,71–76] and on flowers [31,36]. During separate sessions, panelists were also asked to give a personal preference (hedonistic test) to flower samples collected in 2018, although not planned in QDA, in order to assess a preliminary general rating, considering the experience acquired in the previous years [71,73]. Preference was scored from zero (lowest) to 10 (highest), providing judgement for taste, appearance and overall satisfaction.

2.3. Shelf Life

The shelf life evaluation was performed once in 2017 in five species (*C. officinalis, L. vulgare, R. pseudoacacia, S. nigra, T. majus*) and in 2018 in twelve species (*A. ursinum, B. officinalis, C. cyanus, C. intybus, D. carthusianorum, L. angustifolia, P. officinalis, P. veris, R. canina, R. pendulina, S. pratensis, T. officinale*). Five grams of flowers were put into plastic boxes with lid (Ondipack 250 cc, 123 mm × 114 mm × 50 mm, polypropylene, 4.46 g empty, Plemet, France) and stored at 4 °C in a cool chamber (MEDIKA 700, Fiocchetti Cold Manufacturer, Luzzara, Italy) with a transparent glass door, for 14 days, without artificial light. At least five boxes were prepared for each species. The shelf life of flowers was assessed through a visual quality evaluation across the experiment, performed by the same person. A 10-points scale was used, based on visual observation of the degree of decay [21,44] with 10 corresponding to freshly harvested flowers, without imperfections, six was the limit of marketability, while one corresponded to decomposing flowers (wilting, browning).

2.4. Plant Extracts

For each species, three biological replicates of fresh flowers were finely ground with liquid nitrogen at harvest (day 0) and stored at −80 °C until analyses. Then, on days 3, 7, 10, and 14, three biological replicates of fresh flowers were randomly picked from the same plastic boxes used for the shelf life evaluation. The material was ground with liquid nitrogen and stored at −80 °C until analyses. Flowers' extracts were prepared through the ultrasound-assisted extraction method [45,77]. For each sample, 0.5 g of frozen grinded material was put into a glass tube, to which 25 mL of a water:methanol solution (1:1) were added. The tubes were then put into an ultrasound extractor (23 kHz; SARL REUS, Drap, France) for 15 min at room temperature. The extraction procedure was performed once. The phytoextract obtained was filtered through paper filters (Whatman No 1, Whatman, Maidstone, UK) and then maintained at −20 °C until the following analyses.

2.5. Bioactive Compounds

2.5.1. Total Polyphenols

The total phenolic content in the extracts was determined following the Folin–Ciocalteu method [45,78]. The analysis was performed as follows: 1000 µL of diluted (1:10) Folin reagent were mixed with 200 µL of phytoextract in each plastic tube. The samples were left in the dark at room temperature for 10 min, then 800 µL of Na_2CO_3 (7.5%) were added to each tube. Samples were left in the dark at room temperature for 30 min. Absorbance was then measured at 765 nm by means of a spectrophotometer (Cary 60 UV-Vis, Agilent Technologies, Santa Clara, CA, USA), and the results were expressed on a fresh weight basis in milligrams of gallic acid equivalents per 100 g (mg GAE/100 g). The evaluation of total polyphenols was performed in triplicate on extracts of days 0, 3, 7, 10, and 14.

2.5.2. Total Anthocyanins

The total anthocyanins were estimated by pH differential method using two buffer systems: hydrochloric acid/potassium chloride buffer at pH 1.0 (25 mM) and sodium acetate buffer pH 4.5 (0.4 M), as described in the literature [45,79,80]. This method is based on the structural transformation of anthocyanins due to a change in pH (colored at pH 1.0 and colorless at pH 4.5). Briefly, 0.2 mL of each extract was diluted in a 5-mL volumetric flask with the corresponding buffers and the solution was read after 15 min against Milli-Q water as a blank at 510 and 700 nm. By means of a spectrophotometer (Cary 60 UV-Vis, Agilent Technologies, Santa Clara, CA, USA). Absorbance (A) was calculated as follows: A = (A510 nm–A700 nm) pH 1.0 − (A510 nm–A700 nm) pH 4.5. Then, the total anthocyanins (TA) of each extract were calculated by the following equation: $TA = [A \times MW \times DF \times 1000] \times 1/\varepsilon \times 1$, where A is the absorbance; MW is the molecular weight of cyanidin-3-*O*-glucoside (449.2 D); DF is the dilution factor (25); ε is the molar extinction coefficient of cyanidin-3-glucoside (26.900) and results were expressed on a fresh weight basis in milligrams of cyanidin-3-*O*-glucoside per 100 g (mg C3G/100 g). The evaluation of total anthocyanins was performed in triplicate on extracts of days 0, 3, 7, 10, and 14.

2.6. Antioxidant Activity

2.6.1. DPPH Assay

To evaluate the antioxidant activity, the first procedure adopted was the 2,2-diphenyl-1-picrylhydrazyl radical (DPPH) scavenging method [77,81] with slight modifications. The working solution of DPPH radical cation ($DPPH^{\cdot}$, 100 µM) was obtained, dissolving 2 mg of DPPH in 50 mL of MeOH. The solution must have an absorbance of 1.000 (±0.05) at 515 nm. To prepare the samples, 40 µL of phytoextract were mixed with 3 mL of $DPPH^{\cdot}$. Samples were then left in the dark at room temperature for 30 min. Absorbance was measured at 515 nm by means of a spectrophotometer (Cary 60 UV-Vis, Agilent Technologies, Santa Clara, CA, USA). The DPPH radical-scavenging activity was calculated as $[(Abs_0 - Abs_1/Abs_0) \cdot 100]$, where Abs_0 is the absorbance of the control and Abs_1 is the absorbance of the sample. The antioxidant capacity was plotted against a Trolox calibration curve and results were expressed on a fresh weight basis as µmol of Trolox equivalents per gram (µmol TE/g). The DPPH assay was performed in triplicate on extracts of days 0, 3, 7, 10, and 14.

2.6.2. ABTS Assay

The second procedure adopted was the 2,2′-azino-bis(3-ethylbenzthiazoline-6-sulphonic acid) (ABTS) method [79] with slight modification. The working solution of ABTS radical cation ($ABTS^{\cdot}$) was obtained by the reaction of 7.0 mM ABTS stock solution with 2.45 mM potassium persulfate ($K_2S_2O_8$) solution. After the incubation for 12–16 h in the dark at room temperature, the working solution was diluted with distilled water to obtain an absorbance of 0.70 (±0.02) at 734 nm. The antioxidant activity was assessed by mixing 30 µL of phytoextract with 2 mL of diluted $ABTS^{\cdot}$. Samples were left in the dark at room temperature for 10 min. Absorbance was the measured at 734 nm by means of a spectrophotometer

(Cary 60 UV-Vis, Agilent Technologies, Santa Clara, CA, USA). The antioxidant activity was plotted against a Trolox calibration curve and results were expressed on a fresh weight basis as μmol of Trolox equivalents per gram (μmol TE/g). The ABTS assay was performed in triplicate on extracts of days 0, 3, 7, 10, and 14.

2.6.3. FRAP Assay

The third procedure was the FRAP (Ferric ion Reducing Antioxidant Power) method [45,77,82]. The FRAP solution was obtained by mixing a buffer solution at pH 3.6 ($C_2H_3NaO_2$ + $C_2H_4O_2$ in water), 2,4,6-tripyridyltriazine (TPTZ, 10 mM in HCl 40 mM), and $FeCl_3{\cdot}6H_2O$ (20 mM). The antioxidant activity was determined mixing 30 μL of phytoextract with 90 μL of deionized water and 900 μL of FRAP reagent. The samples were then placed at 37 °C for 30 min. Absorbance was measured at 595 nm by means of a spectrophotometer (Cary 60 UV-Vis, Agilent Technologies, Santa Clara, CA, USA). Results were expressed on a fresh weight basis as mill moles of ferrous iron equivalents per kilogram (mmol Fe^{2+}/kg). The FRAP assay was performed in triplicate on extracts of days 0, 3, 7, 10, and 14.

2.7. Statistical Analysis

Data of the sensory profiles, visual quality, bioactive compounds, and antioxidant activity were previously subjected to the Shapiro–Wilk normality test and Levene homogeneity test ($p > 0.05$). The differences between species and across time were computed using a parametric or a non-parametric one-way analysis of variance (ANOVA), according to the significance of the previous tests and means were separated with Tukey's HSD test ($p \leqslant 0.05$). The Pearson's correlation was performed on QDA values and subjective judgement. The Spearman's correlation was performed on polyphenols, anthocyanins, DPPH, ABTS, FRAP and sensory profile values. The principal component analysis (PCA) was performed on sensory data to visualise the contribution of each attribute to the overall variability. The partial least square (PLS) regression was also done to investigate correlations between phytochemical profile (X-variables) and sensory data (Y-variables) after standardization of the data; the phytochemical profile of the 17 species derives from recent work on the same plant material [13]. All data were analyzed by means of the SPSS software (version 25.0; SPSS Inc., Chicagom, IL, USA), except for PCA (Past 4.01, [83]), and spider charts were prepared using Microsoft Office Excel.

3. Results

3.1. Sensory Analysis

3.1.1. Lexicon and QDA Sensory Sheet Definition

The 10 selected sensory descriptors are defined in Table 2, with four descriptors for taste (sweet, sour, bitter, and salt), three for aroma (smell, specific flower aroma, and herbaceous aroma), together with chewiness, astringency and spiciness. A specific sensory analysis sheet for flower evaluations was realized (Figure S1) and used for the QDA test, using a reduced list of sensory lexicon both to ease the judges' evaluation and to describe the essential traits of samples.

3.1.2. Sensory Profiles

The detailed sensory profiles of the 17 species are shown in Table 3 and Figures S2–S4. A wide variability was recorded among the tested edible flowers in terms of the range of intensities, with spiciness having the widest range of variation (7.4), followed by specific flower aroma (6.2), bitterness and sweetness (6.1), smell (6.1), herbaceous aroma (4.7), chewiness (4.4), astringency (4.1), sour intensity (2.5), and salt intensity as the least variable descriptor (2.4). All the sensory descriptors were detected in each flower species, except for spiciness that was absent in *R. pseudoacacia*. The highest intensities were recorded for smell in *L. angustifolia* (9.0) and specific flower aroma in *A. ursinum* (8.8).

Table 2. Sensory lexicon used in this study: descriptors, definitions and bibliographic references.

Sensory Descriptor	Definition	References
Sweet intensity	Taste of sucrose	[84–86]
Sour intensity	Taste of citric acid	[85,87,88]
Bitter intensity	Taste of caffeine	[85,89]
Salt intensity	Taste of sodium chloride	[85,88]
Smell intensity	Odor's intensity of edible flower in evaluation	[87,90]
Specific flower aroma intensity	Aroma's intensity of edible flower in evaluation	[2,9,87,90,91]
Herbaceous aroma intensity	Intensity of herbaceous and cut grass aroma	[87]
Spiciness	Intensity of spice aroma, hot and pungent taste	[85,87,92,93]
Chewiness	The amount of chewing required to break down the sample so that it can be swallowed	[88]
Astringency	The tactile sensation described as dryness, tightening, tannic and puckering sensations perceived in the oral cavity.	[94–96]

Table 3. Intensities (from 0 to 10) of each sensory descriptor detected in the studied edible flowers.

	Smell	Sweet	Sour	Bitter	Salt	Specific Flower Aroma	Herbaceous Aroma	Spiciness	Chewiness	Astringency
Allium ursinum	8.3	2.4	1.8	2.1	2.7	8.8	1.4	6.1	7.4	0.3
Borago officinalis	4.3	3.6	0.8	1.2	0.8	3.9	2.4	0.1	6.1	0.4
Calendula officinalis	6.7	2.6	1.4	3.0	0.8	5.2	2.2	0.6	7.2	2.1
Centaurea cyanus	5.1	2.2	1.1	2.2	0.9	4.1	3.4	0.2	3.8	0.7
Cichorium intybus	3.1	0.9	2.8	7.2	0.7	5.5	4.1	1.3	5.8	1.3
Dianthus carthusianorum	6.7	1.9	0.4	2.9	0.5	6.0	1.3	1.4	4.6	0.5
Lavandula angustifolia	9.0	2.8	1.7	5.0	0.5	8.2	2.5	1.8	4.7	0.7
Leucanthemum vulgare	7.4	2.4	0.9	2.7	0.9	3.1	5.3	0.7	5.2	1.5
Paeonia officinalis	4.1	3.9	2.9	6.1	0.6	5.1	5.1	1.6	7.9	1.9
Primula veris	4.1	2.8	0.8	1.2	0.4	2.6	1.6	0.3	6.2	1.4
Robina pseudoacacia	7.1	6.9	1.0	1.4	0.9	5.9	3.1	-	6.3	1.2
Rosa canina	6.3	2.0	2.0	5.2	0.5	6.1	4.2	0.1	6.3	2.0
Rosa pendulina	5.5	1.3	2.1	7.3	0.3	5.0	2.5	0.6	6.5	4.4
Salvia pratensis	7.3	3.9	1.1	2.0	0.6	5.3	2.0	0.8	6.6	0.8
Sambucus nigra	7.8	3.5	0.7	3.5	1.3	6.7	2.8	1.3	7.4	0.9
Taraxacum officinale	6.4	3.7	0.6	1.4	0.8	4.5	1.4	0.4	5.7	0.4
Tropaeolum majus	8.3	0.8	0.8	5.4	1.6	7.3	0.6	7.4	8.2	1.9
Range of variation	5.9	6.1	2.5	6.1	2.4	6.2	4.7	7.4	4.4	4.1

In *A. ursinum* (Table 3, Figure S2), the intensities of smell and garlic aroma were very high (8.3 and 8.8, respectively) and flowers were easy to chew (7.4). *Borago officinalis* had a marked chewiness (6.1), but was not astringent neither spicy and taste descriptors (sweet, sour, bitter and salt) were lower than 3.5. *Calendula officinalis* had medium smell (6.7) and aroma (5.2), not very marked for taste but easy to chew and little astringent and spicy. The QDA profile of *C. cyanus* had low values, with the most marked descriptors (smell, aroma, and chewiness) lower than 6. The sensory profile of *C. intybus* was defined by bitter taste, aroma intensity and chewiness.

The flowers of *D. carthusianorum* (Table 3, Figure S3) had 6.7 of smell and 6 of aroma, while bitterness was the most marked of the taste descriptors. The profile of *L. angustifolia* had very high values for smell and aroma intensity (9.0–8.2); the chewiness was medium (4.7) and bitterness characterized the taste. *Leucanthemum vulgare* had values higher than 5 only in smell (7.4), herbaceous aroma (5.3), and chewiness (5.2). The panel scored a high chewiness for *P. officinalis* and among the taste descriptors, the bitter taste was the highest (6.1).

Primula veris had values lower than 5 in all descriptors, except for chewiness (6.3), so the organoleptic sensations are delicate. The sensory profile of *R. pseudoacacia* showed high values for smell (7.0), sweet (7.0), and aroma (6.0), and chewiness was sufficiently easy (6.0).

The profile of *R. canina* (Table 3, Figure S4) reveals an easy chewiness (6.6) and bitterness was the most marked of the taste descriptors (4.8). *Rosa pendulina* flowers had higher bitter taste (7.2) and astringency (4.4) and lesser herbaceous aroma (2.5) than *R. canina* flowers. *Salvia pratensis* was chewable (6.6), with high smell (7.3) and medium aroma (5.3); all the other descriptors were equal or lower than 2.0, except for the sweet taste (3.9). *Sambucus nigra* had a sensory profile with high intensities of smell (7.8) and aroma (6.7); it was easy to chew (7.4), sweet and bitter intensities were balanced (3.5), while all the other descriptors were lower than 2. In *T. officinale*, the smell and aroma were sufficiently marked (6.4 and 4.5 respectively), as per chewiness (5.7), while sweet intensity was the highest among the taste descriptors, though being 3.7. Finally, *T. majus* had very high intensities of smell (8.3) and aroma (7.3) and was easy to chew (8.2), spicy (8.2), and quite bitter (5.4), while all the other descriptors were lower than 2.

The PCA plot generated from the sensory data is shown in Figure 2. Component 1 (PC1) accounted for 25.4% of the sensory variation in the studied edible flowers and PC2 accounted for 20.0%, explaining 45.4% of sensory descriptors variability. PC1 has a positive association mostly with the variation of specific flower aroma and spiciness, with *T. majus* and *A. ursinum* showing the highest values; while PC2 mainly reflected sour, bitter, and herbaceous aroma intensities with *C. intybus*, *P. officinalis*, and *R. pendulina* showing the highest values. The other species are scattered on the plot, with intermediate or negative relation with most of the sensory descriptors, except for sweet intensity, which characterizes *S. pratensis*, *D. carthusianorum*, *R. pseudoacacia*, and *T. officinale*. According to the loadings, sour and bitter intensity and astringency (upper right quadrant) are inversely related with sweet intensity (lower left quadrant), while to a lesser extent, smell intensity (lower right quadrant) is inversely related with herbaceous aroma intensity (upper right quadrant).

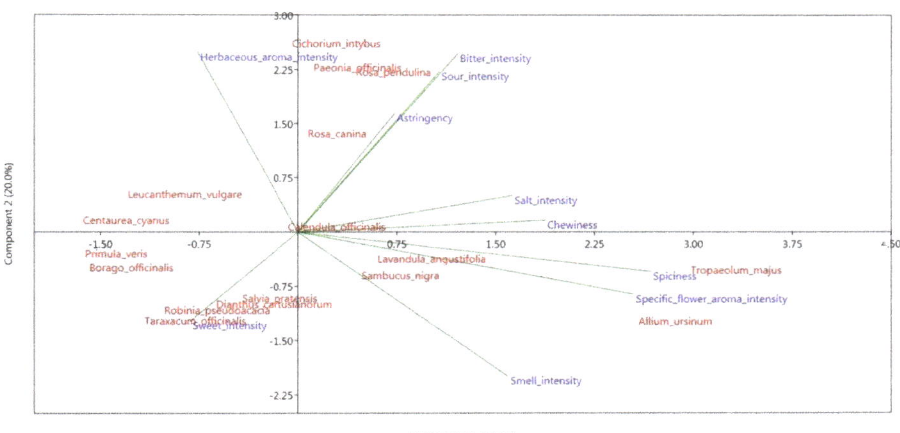

Figure 2. PCA biplot of the sensory descriptors of 17 edible flowers.

The data acquired on the sensory traits of the 17 species were evaluated together with the HPLC-DAD phytochemical profiles of the same plant material reported in a recent work [13]. In particular, sensory data were correlated with the content of phenolic acids (cinnamic acids: caffeic, chlorogenic, coumaric and ferulic acid; benzoic acids: ellagic and gallic acid), flavonols (hyperoside, isoquercitrin, quercetin, quercitrin and rutin), flavanols (catechin and epicatechin), and vitamin C with PLS regression. Cumulative Q^2 of component 1 (0.059) and component 2 (0.163) and cumulative R^2 of both X and Y in the two components were below 0.3 (R^2Y comp.1 = 0.099, R^2Y comp.2 = 0.223, R^2X comp.1 = 0.167, R^2X comp.2 = 0.274) suggesting weak relations between descriptors and compounds.

3.1.3. Subjective Judgement

Concerning the satisfaction rating (Table 4), the flowers showed very different levels of acceptance. The overall subjective judgement ranged from 4.63 of *C. intybus* to 7.07 of *A. ursinum*. Regarding taste, the subjective judgement of panel members ranged from 3.25 (*R. pendulina*) to 6.57 (*A. ursinum*). Regarding appearance, the different species were generally appreciated (from 6.21 in *T. officinale* to 7.64 in *P. officinalis*) except for *C. intybus* (3.25) and *D. carthusianorum* (4.60).

Table 4. Subjective judgement (0–10 of satisfaction rating) on edible flower.

Species	Overall	Taste	Appearance
Allium ursinum	7.07 ± 0.93 a [1]	6.57 ± 0.79 a	7.21 ± 0.91 a
Borago officinalis	5.60 ± 0.55 ab	4.60 ± 0.55 abcde	6.60 ± 0.55 a
Centaurea cyanus	5.64 ± 0.84 ab	4.25 ± 0.94 bcde	7.32 ± 0.87 a
Cichorium intybus	4.63 ± 0.75 b	4.00 ± 0.82 cde	3.25 ± 0.96 c
Dianthus carthusianorum	4.80 ± 0.45 b	4.20 ± 0.84 bcde	4.60 ± 0.89 bc
Lavandula angustifolia	6.30 ± 0.84 ab	5.30 ± 0.97 abcd	7.60 ± 0.55 a
Paeonia officinalis	6.93 ± 0.93 a	6.21 ± 0.99 ab	7.64 ± 0.99 a
Primula veris	5.43 ± 0.98 ab	3.29 ± 0.95 de	6.36 ± 0.99 ab
Rosa canina	5.57 ± 0.98 ab	4.43 ± 0.79 bcde	7.36 ± 0.63 a
Rosa pendulina	5.25 ± 0.50 ab	3.25 ± 0.96 e	7.50 ± 0.71 a
Salvia pratensis	6.00 ± 0.82 ab	5.25 ± 0.50 abcde	6.50 ± 0.58 ab
Taraxacum officinale	6.00 ± 0.99 ab	5.86 ± 0.90 abc	6.21 ± 0.99 ab

[1] Mean value ± standard deviation of each sample is given. Values with the same letter within the same column are not statistically different ($p < 0.01$) according to Tukey's HSD test.

Despite the positive significant correlations found between the overall subjective judgement and the intensity of smell, sweet and aroma of specific flower (Table 5, $p < 0.01$), and the intensity of herbaceous aroma ($p < 0.05$), these are often weak (below 0.45). The correlation between the overall judgment and salt intensity ($p < 0.01$) and astringency was instead negative ($p < 0.05$). The overall subjective judgement was not significantly correlated with sour and bitter intensity, spicy and chewiness.

Table 5. Pearson's correlation between sensory parameters and overall subjective judgement.

Sensory Descriptors	Overall Subjective Judgement	Pearson Correlation Significance
Smell intensity	0.342	** [1]
Sweet intensity	0.421	**
Sour intensity	−0.009	ns
Bitter intensity	−0.135	ns
Salt intensity	−0.234	**
Specific flower aroma intensity	0.272	**
Herbaceous aroma intensity	0.179	*
Spicy	0.510	ns
Chewiness	0.022	ns
Astringency	−0.171	*

[1] the level of significance is given: *, $p < 0.05$; **, $p < 0.01$; ns, not significant.

3.2. *Bioactive Compounds and Antioxidant Activity at Harvest*

Bioactive compounds and antioxidant activity of freshly harvested flowers are reported in Figures 3 and 4. Particularly, polyphenols (Figure 3A) ranged from 76.41 mg GAE/100 g FW (*T. officinale*) and 1270.72 mg GAE/100 g FW (*P. officinalis*). Anthocyanins (Figure 3B) ranged from 0.58 mg C3G/100 g FW (*A. ursinum*) and 800.23 mg C3G/100 g FW of *T. majus*, which had four times higher values than the second species in ranking (*S. pratensis*). The antioxidant activity, measured through different assays, had a similar pattern in DPPH and ABTS methods (Figure 4A,B), while the FRAP differed (Figure 4C). DPPH values ranged from 2.08 µmol TE/g FW (*R. pseudoacacia*) and 232.44 µmol TE/g FW (*P. officinalis*). Both roses had high DPPH scavenging activity (153.96 µmol TE/g FW in *R. pendulina* and 132.25 µmol TE/g FW in *R. canina*), followed by *C. intybus* (69.17 µmol TE/g FW) and all the other species. Concerning ABTS, values ranged from 2.70 µmol TE/g FW (*A. ursinum*) to 57.59 µmol TE/g FW (*P. officinalis*). As for the DPPH assay, roses had high scavenging activity (55.44 µmol TE/g FW in *R. pendulina* and 43.45 µmol TE/g FW in *R. canina*), followed by *C. intybus* (26.85 µmol TE/g FW) and all the other species. FRAP values (Figure 4C) ranged from 1.45 mmol Fe^{2+}/kg FW (*A. ursinum*) to 274.22 mmol Fe^{2+}/kg FW (*P. officinalis*). In this assay, *T. majus* showed very high antioxidant activity (241.12 mmol Fe^{2+}/kg FW), comparable to that of *R. canina* (203.72 mmol Fe^{2+}/kg FW), *R. pendulina* (257.04 mmol Fe^{2+}/kg FW), *S. pratensis* (171.09 mmol Fe^{2+}/kg FW), *C. intybus* (138.36 mmol Fe^{2+}/kg FW), and *P. veris* (120.14 mmol Fe^{2+}/kg FW).

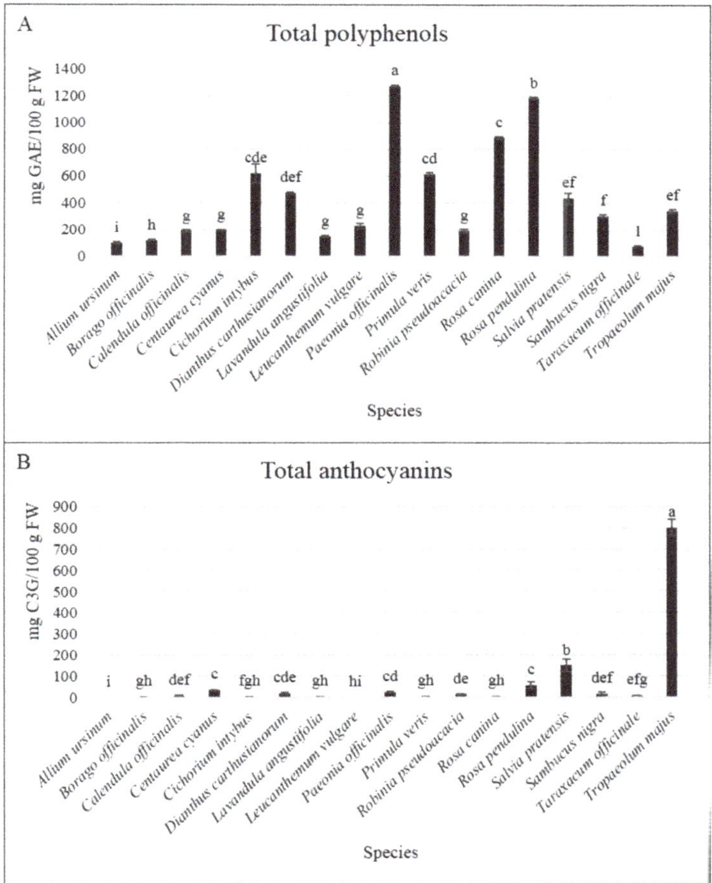

Figure 3. Total phenolic content (**A**) and total anthocyanin content (**B**) of fresh flowers at harvest (day 0) in all the analyzed species. Data are given as mean values; bars indicate standard error. Different letters correspond to significant differences between means according to Tukey's HSD test ($p < 0.05$).

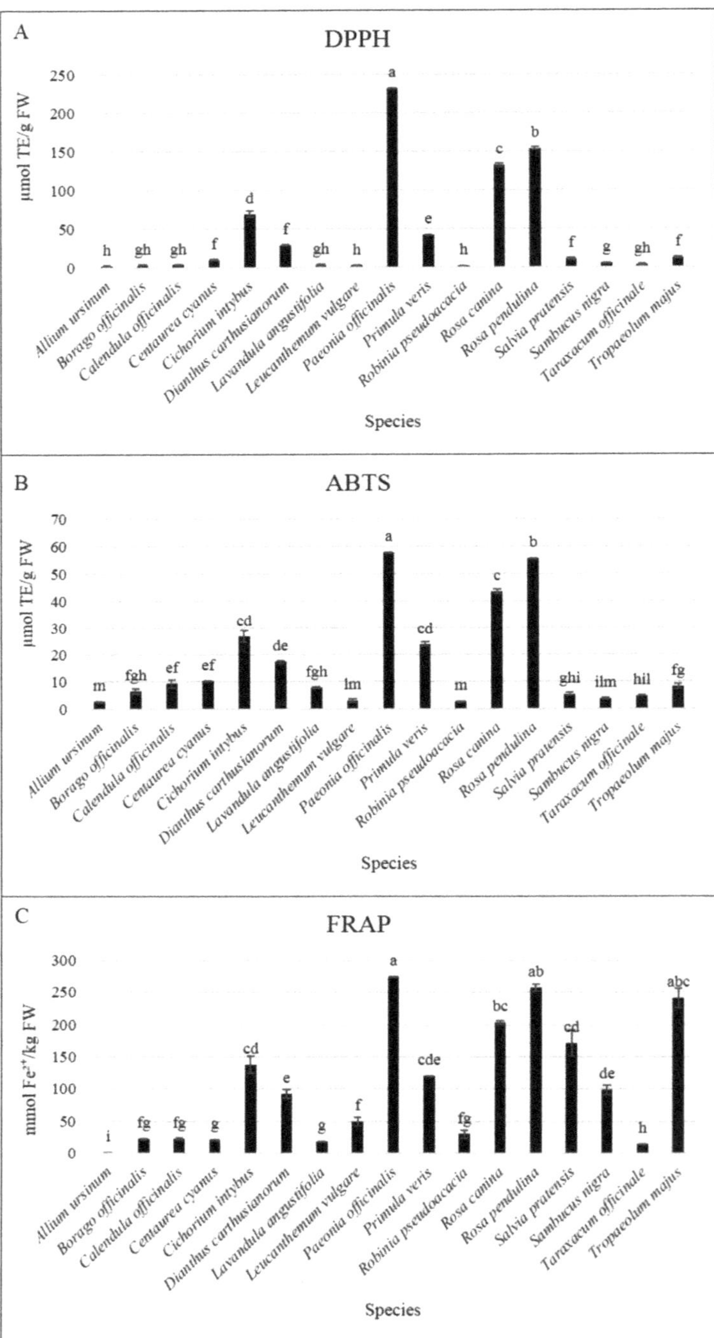

Figure 4. Antioxidant activity of fresh flowers at harvest in all the analyzed species, according to (**A**) DPPH, (**B**) ABTS, and (**C**) FRAP assay. Data are given as mean values; bars indicate standard error. Different letters correspond to significant differences between means according to Tukey's HSD test ($p < 0.05$).

The correlation analysis (Table 6) between the content of polyphenols and anthocyanins at harvest and the antioxidant activity evaluated through different assays indicated that all these parameters are significantly correlated, except for the content of anthocyanins and ABTS values (p = 0.091). All the correlations were positive and the three methods of analysis for the antioxidant activity were highly related. The total polyphenol and anthocyanin content at harvest was also evaluated in relation to the sensory profiles of the species. Few correlations were recorded, being both groups of bioactive compounds negatively correlated only with salt intensity (polyphenols: $r = -0.275$, $p < 0.001$; anthocyanins: $r = -0.299$, $p < 0.001$).

Table 6. Spearman's correlation coefficient (r) and related level of significance between polyphenols, anthocyanins, and antioxidant activity measured with DPPH, ABTS, and FRAP assays.

		Polyphenols	**Anthocyanins**	**DPPH**	**ABTS**	**FRAP**
Polyphenols	r	1	0.270	0.813	0.649	0.895
	Sign.		0.032	0.000	0.000	0.000
Anthocyanins	r		1	0.386	0.214	0.444
	Sign.			0.002	0.091	0.000
DPPH	r			1	0.837	0.793
	Sign.				0.000	0.000
ABTS	r				1	0.557
	Sign.					0.000
FRAP	r					1
	Sign.					

3.3. Shelf Life and Dynamics of Bioactive Compounds

Visual quality grade and the content of total polyphenols and anthocyanins in the studied species are reported in Table 7. Data on *C. intybus* was not available, since the flowers rapidly rotted and no further evaluations were possible beyond day 0. Roses had acceptable visual quality rate (⩾6) for the longest period, up to day 10 in *R. canina* and up to day 14 in *R. pendulina*. All the other species lasted seven days, while *T. officinale* did not reach day 3.

Table 7. Visual quality of fresh edible flowers during cold storage at 4 °C (0, 3, 7, 10, 14 days after harvest). Data shown are mean values.

Days	*A. ursinum*		*D. carthusianorum*		*P. veris*		*S. pratensis*	
0	10	a [1]	10	a	10	a	10	a
3	8.5	b	8.3	b	6.5	b	9	b
7	6.5	c	7.2	c	5.6	bc	8	c
10	5.8	d	5	d	5	c	4.2	d
14	5	e	3	e	4.8	c	3.7	e
	***		***		***		***	
	B. officinalis		*L. angustifolia*		*R. pseudoacacia*		*S. nigra*	
0	10	a	10	a	10	a	10	a
3	9	ab	8	b	7.3	b	9	b
7	8.2	b	7.2	c	7.3	b	8.8	b
10	5.4	c	5.2	d	4.8	c	4	c
14	4.6	c	4	e	3.8	d	2.5	d
	***		***		***		***	
	C. officinalis		*L. vulgare*		*R. canina*		*T. officinale*	
0	10	a	10	a	10	a	10	a
3	8.6	b	8.1	b	8.8	b	3.3	b
7	7	c	7.7	b	8.3	b	2	c
10	4.5	d	5.9	c	7.3	c	2	c
14	1.7	e	5.3	c	5.8	d	1	d
	***		***		***		***	

Table 7. *Cont.*

	C. cyanus		*P. officinalis*		*R. pendulina*		*T. majus*	
0	10	a	10	a	10	a	10	a
3	9	b	8.4	b	9	b	8.6	b
7	8.8	b	7.8	b	6.8	c	6	c
10	5.6	c	5.7	c	6.7	cd	4.7	d
14	4.6	d	4.7	c	5.8	d	1	d
	***		***		***		***	

[1] Data with different letters are significantly different according to Tukey's HSD test; the level of significance is given (***, $p < 0.001$).

Concerning the bioactive compounds, the total phenolic content (Table 8) during storage varied, increasing significantly in eight species (*B. officinalis*, *C. cyanus*, *L. angustifolia*, *L. vulgare*, *P. veris*, *R. canina*, *t. officinale*) and decreasing in 4 (*A. ursinum*, *C. officinalis*, *P. officinalis*, *T. majus*) while it remained stable in *R. pseudoacacia*, *R, pendulina*, *S. pratensis*, and *S. nigra*. The total anthocyanin content varied to a lesser extent (Table 8), as only two species showed significant variation across the trial, namely *T. officinale* and *T. majus*. Whereas all the other species had constant values of anthocyanins during storage.

Table 8. Total polyphenols and total anthocyanins of fresh edible flowers during cold storage at 4 °C (0, 3, 7, 10, 14 days after harvest). Data shown are mean values expressed on a fresh weight (FW) basis.

Days	Total Polyphenols mgGAE/100 g FW		Total Anthocyanins mg C3G/100 g FW		Total Polyphenols mg GAE/100 g FW		Total Anthocyanins mg C3G/100 g FW		Total Polyphenols mg GAE/100 g FW		Total Anthocyanins mg C3G/100 g FW	
	A. ursinum				*L. vulgare*				*R. pendulina*			
0	99.26	ab [1]	0.58		230.76	b	0.83		1181.9	a	55.93	
3	109.2	ab	0.77		197.91	b	4.3		1043.5	bc	47.22	
7	138.7	a	1.4		338.45	a	6.67		1151.4	ab	60.13	
10	81.51	b	0.23		313.52	ab	4.32		1033.9	c	31.03	
14	47.96	c	0.92		343.48	a	13.93		1139.8	abc	39.48	
	*		ns		*		ns		*		ns	
	B. officinalis				*P. officinalis*				*S. pratensis*			
0	118.1	b	38.43		1270.7	a	25.96		433.93		150.7	
3	100.8	b	40.27		1210.9	b	28.76		425.35		135.5	
7	100.6	b	3.89		1208.6	b	25.58		396.64		107.4	
10	115.6	b	45.43		1219.8	b	26.68		349.52		141.8	
14	157.2	a	44.73		1161.1	c	26.2		457.66		118.7	
	***		ns		***		ns		ns		ns	
	C. officinalis				*P. veris*				*S. nigra*			
0	189.6	a	9.9		609.16	b	3.99		307.64	ab	17.29	
3	156	ab	35.66		770.23	a	4.24		365.14	a	29.46	
7	149.2	ab	26.46		610.58	b	4.86		284.33	b	16.03	
10	136.2	b	20.11		638.24	b	4.96		315.84	ab	15.22	
14	135.7	b	19.61		731.4	a	3.52		292.16	b	11.72	
	*		ns		***		ns		**		ns	
	C. cyanus				*R. pseudoacacia*				*T. officinale*			
0	196.8	d	34.67		191.45		14.45		76.41	d	8.84	ab
3	171.5	e	50.12		204.08		9.84		157.33	a	6.05	b
7	276	b	28.57		243.17		10.83		100.35	c	10.62	a
10	213.5	c	23.11		187.53		15.49		117.89	b	8.68	ab
14	317.9	a	38.04		205.6		11.95		98.62	c	8.57	ab
	***		ns		ns		ns		***		**	

Table 8. *Cont.*

Days	Total Polyphenols mgGAE/100 g FW		Total Anthocyanins mg C3G/100 g FW	Total Polyphenols mg GAE/100 g FW		Total Anthocyanins mg C3G/100 g FW	Total Polyphenols mg GAE/100 g FW		Total Anthocyanins mg C3G/100 g FW	
	D. carthusianorum			*R. canina*			*T. majus*			
0	470.5	c	19.16	884.44	d	4.39	341.33	a	800.2	a
3	446.8	c	15.92	1009.6	c	5.04	343.64	a	414.2	b
7	675.9	a	17.25	1204.2	a	4.96	353.95	a	327.5	b
10	635.9	b	16.49	1155	ab	4.5	271.93	a	335.8	b
14	697.7	a	18.97	1104.6	b	5.2	48.74	b	322	b
	***		ns	***		ns	***		**	
	L. angustifolia									
0	148.2	c	4.27							
3	198.7	bc	4.94							
7	202.9	bc	5.03							
10	212.3	b	4.98							
14	387.5	a	5							
	*		ns							

[1] Data with different letters are significantly different according to Tukey's HSD test; the level of significance is given (*, $p < 0.05$; **, $p < 0.01$; ***, $p < 0.001$; ns, not significant).

Antioxidant Activity during Postharvest

The antioxidant activity measured with the DPPH assay significantly increased in four species (*A. ursinum*, *D. carthusianorum*, *L. angustifolia*, *L. vulgare*), decreased in four species (*C. officinalis*, *P. officinalis*, *R. canina*, *R. pendulina*) and remained constant in the others during 14 days of storage (Table 9).

Table 9. Antioxidant activity (DPPH, ABTS, FRAP assays) of fresh edible flowers during cold storage at 4 °C (0, 3, 7, 10, 14 days after harvest). Data shown are mean values expressed on a fresh weight (FW) basis.

Days	DPPH µmol TE/g FW		ABTS µmol TE/g FW		FRAP mmol Fe^{2+}/kg FW		DPPH µmol TE/g FW		ABTS µmol TE/g FW		FRAP mmol Fe^{2+}/kg FW	
			A. ursinum						*P. veris*			
0	2.38	c [1]	2.7	c	1.45	b	41.91	abc	23.83	bc	120.14	b
3	6.76	a	7.41	a	4.01	a	42.63	ab	24.7	bc	115.17	b
7	4.75	b	3.32	b	5.04	a	53.29	a	22.96	c	114.25	b
10	4.73	b	5.56	ab	3.77	a	28.68	bc	27.67	a	118.84	b
14	4.79	b	4.65	ab	4.22	a	29.96	c	25.43	ab	131.17	a
	***		*		***		*		***		**	
			B. officinalis						*R. pseudoacacia*			
0	3.47	a	6.53		22.74	b	2.08	ab	2.66	ab	30.35	ab
3	1.81	b	4.61		15.63	d	2.53	a	3.43	a	40.44	a
7	1.28	b	5.34		18.06	c	1.86	ab	4.16	a	47.1	a
10	4.59	a	7.92		24.19	b	1.82	b	3.4	ab	24.66	b
14	4.8	a	8.16		33.56	a	1.4	b	2.56	b	18.83	c
	***		ns		***		**		*		**	
			C. officinalis						*R. canina*			
0	3.62	a	9.21		22.55	b	132.3	b	43.45	c	203.72	b
3	1.29	bc	2.39		34.16	a	137.7	b	52.86	b	227.37	ab
7	1.06	bc	2.3		32.56	a	137.4	b	50.91	bc	239.1	ab
10	0.97	c	2.34		31.34	a	177.6	a	57.39	a	265.09	a
14	1.89	b	1.97		34.87	a	115.8	c	49.98	bc	248	ab
	***		ns		***		***		*		*	

Table 9. *Cont.*

Days	DPPH µmol TE/g FW		ABTS µmol TE/g FW		FRAP mmol Fe^{2+}/kg FW		DPPH µmol TE/g FW		ABTS µmol TE/g FW		FRAP mmol Fe^{2+}/kg FW	
			C. cyanus						*R. pendulina*			
0	10.08	b	10.3	b	21.19	c	153.9	b	55.44	ab	257.04	
3	10.77	b	13.4	ab	34.31	bc	104.8	c	55.35	ab	248.22	
7	16.04	a	16.9	a	40.29	b	170.6	a	56.29	a	246.11	
10	11.59	b	14	ab	26.57	bc	99.85	d	54.62	b	251.63	
14	12.67	ab	13.7	ab	44.65	a	89.77	e	56.36	a	224.1	
	**		*		*		**		**		ns	
			D. carthusianorum						*S. pratensis*			
0	29.13	c	17.4	b	92.51	b	11.02		5.39		171.09	
3	17.19	d	12.8	c	69.6	b	10.78		5.23		151.51	
7	37.59	b	20	b	121.3	a	10.3		5.64		144.54	
10	34.87	bc	27.1	a	127.8	a	10.62		5.26		170.26	
14	58.3	a	25.5	a	131.4	a	11.76		5.65		194.11	
	***		***		***		ns		ns		ns	
			L. angustifolia						*S. nigra*			
0	2.78	bc	7.77	b	19.25	c	5.14	ab	3.74	b	98.79	
3	9.72	b	8.75	ab	32.97	abc	6.64	a	5.4	a	109.38	
7	4.19	bc	8.9	ab	36.12	ab	4.34	b	4.35	ab	93.79	
10	3.3	c	8.97	ab	31.14	bc	5.35	ab	4.24	ab	92.91	
14	25.05	a	16.7	a	71.61	a	4.14	b	3.76	b	83.46	
	*		*		*		**		*		ns	
			L. vulgare						*T. officinale*			
0	2.49	cd	3.08		50.08	b	3.18	b	4.78		14.41	bc
3	1.85	d	2.35		49.49	b	6.87	a	7.06		18.46	ab
7	3.86	bc	3.4		77.88	a	2.8	b	6.39		12.15	c
10	5.54	ab	3.18		89.05	a	9.59	a	6.16		22.29	a
14	5.74	a	3.51		96.72	a	2.53	b	6.53		15.59	bc
	***		ns		***		***		ns		***	
			P. officinalis						*T. majus*			
0	232.4	a	57.6	b	274.2	a	11.51	a	8.05		241.12	a
3	227.3	b	57.9	a	265.8	b	5.85	b	4.73		129.15	b
7	190.9	c	57.5	b	275	a	5.62	b	5.02		113.86	b
10	221	c	57.6	b	274.3	a	4.53	b	5.69		119.76	b
14	187.9	c	55.7	c	261.1	b	6.79	ab	5.88		83.47	b
	*		*		***		*		ns		***	

[1] Data with different letters are significantly different according to Tukey's HSD test; the level of significance is given (*, $p < 0.05$; **, $p < 0.01$; ***, $p < 0.001$; ns, not significant).

The antioxidant activity measured with the ABTS assay increased in 3 species (*A. ursinum*, *D. carthusianorum*, *L. angustifolia*) across the trial, decreased in *P. officinalis* and remained constant in the others.

Finally, the antioxidant activity measured with the FRAP assay throughout the trial increased in eight species (*A. ursinum*, *B. officinalis*, *C. officinalis*, *C. cyanus*, *D. carthusianorum*, *L. angustifolia*, *L. vulgare*, *P. veris*), decreased in three species (*P. officinalis*, *R. pseudoacacia*, *T. majus*) and remained constant in the others.

4. Discussion

4.1. Sensory Evaluation

In the present study, trained panelists have described the sensory profiles of 17 edible flower species. Sensory analyses or hedonistic evaluations concerning edible flowers are few; previous studies concerned garlic, pansies, borage, calendula, and nasturtium, among the others [8,27,29,31–36]. Since sensory science is still evolving on these emergent food products, a common lexicon of organoleptic terminology was guaranteed by comparison with specific literature (Table 2), studying and preparing a sensory sheet with a reduced list of sensory descriptors to ease the judges' evaluation and to describe the essential traits of the samples.

The detected sensory profiles were very different from each other in terms of the intensity of each descriptor, highlighting peculiar traits. *Allium ursinum*, *D. carthusianorum*, *L. angustifolia*, and *L. vulgare* were featured by a strong aroma (smell, herbaceous and specific flower aroma) but a poor taste (sweet, sour, bitter, and salt), while in *C. intybus*, *P. officinalis*, *R. pseudoacacia*, and *R. pendulina*, one of the taste descriptor overcame the aromatic traits. The other species (*B. officinalis*, *C. officinalis*, *C. cyanus*, *P. veris*, *R. canina*, *S. pratensis*. *T. officinale*, and *T. majus*) had medium or quite high aroma, and one predominant taste. Among these species, *B. officinalis* resulted in a poor sensory profile, confirming previous results [29], while other authors [31] conferred a very high score to sweet taste of *B. officinalis*, which resembled cucumber, despite not being very fragrant. *Calendula officinalis* also had a poor sensory profile, but an easy chewiness, which are in contrast with a previous report [31], where this flower had a notably bitter taste, not easy chewiness and an affinity to saffron taste.

Spiciness was the most variable descriptor out of the 10 considered, characterizing the profile of *A. ursinum* and *T. majus*, as recorded in previous reports [31,36].

Sensory analysis is essential to understand the potential and most suitable use and combination of each edible flower in the food industry. Differences between species in taste and aroma are ascribed to the different chemical composition of each flower, constituted by hundreds of compounds [14]. The fruity and floral aromas in flowers are, for example, due to the presence of volatile compounds such as ethyl octanoate, 1-hexanol, ρ-cymene, or β-myrecene [29], and flavonoids are responsible for the astringent and bitter taste of foods [97], while organic acids confer sour taste [98]. In this study, few correlations were found between sensory descriptors and higher presence of bioactive compounds (polyphenols and anthocyanins), except for a decreased intensity of salted taste. The multivariate analyses (e.g., PCA or PLS regression) are increasingly used in the sensory science in the attempt of highlighting the degree of contribution of each interdependent sensory attribute to the overall variability of data, or assessing the correlation between the sensory attributes and the analytical data [99–104].

In this study, the PCA output confirmed the wide variability of the species in terms of sensory traits, highlighting interesting taste and aroma intensities. However, the Q^2 and R^2 coefficients of PLS regression suggested that the current model does not fully elucidate the role of phytochemical compounds in the sensory profiles of the studied edible flowers. Despite the importance of bioactive compounds in human nutrition, they are still of difficult sensory perception and further studies are needed to understand which compounds are responsible for the taste and aroma of flowers. The phytochemical profile of edible flowers is affected by environmental and agronomic conditions [105,106] and it is of major importance to standardize the cultivation of each species in order to obtain a uniform food produce not only in terms of appearance, but also in terms of sensory profile, both highly affecting consumer preferences [2,35,36]. This would lead to better satisfy consumers, which are almost unaware of flowers as a food product, but are curious and willing to eat them [2,8,14,27,32–35].

4.2. Shelf Life

Fresh edible flowers are commonly considered highly perishable products, as they rapidly decay within few days after harvest [11,37,38]. Each species has different storage requirements [2], according to their moisture content and respiration rate [21]. Few studies examined the storage conditions of edible flowers, receiving much less attention than cut flowers, vegetables and fruits. An increasing number of trials are thus necessary to understand their postharvest requirements [11,37,38,46,107].

Thirteen species out of the 17 analyzed in this study were acceptable for seven days if stored at 4 °C in polypropylene boxes, agreeing with several reports that mostly indicated the limit of acceptance within one week of cold storage [21,37,38,45,107]. Among these, *B. officinalis*, which showed a seven-day shelf life, also spoiled when stored at 0, 2.5, or 5 °C in sealed polyethylene film bags [38], despite the orage flowers lasting two weeks at −2.5 °C [38] or only one day at 4 °C in another study [107].

Centaurea cyanus also lasted seven days; however, showed a satisfactorily shelf life for 12 days in another study [107]. *D. carthusianorum* showed an acceptable quality for seven days, similar to the more common edible carnation (*Dianthus caryophyllus* L.) in commercial packaging at 5 °C [107]. *Salvia pratensis* (seven-day shelf life) lasted more than the common sage (*Salvia discolor* Kunth, five days [21]), and *Salvia* hybrid (six days at 5 °C in controlled permeability films with 14 h of light [108]). *Tropaeolum majus* was acceptable for seven days, as observed also by [38] at 5 °C, but this flower could last two weeks if stored in the dark, in sealed polyethylene film bags at 0 and 2.5 °C [38]. The species with the longest shelf life were roses, as *R. canina* lasted 10 days and *R. pendulina* 14 days, being the most suitable for sale. Conversely, *C. intybus* and *T. officinale* were the least interesting products, not suitable for storage using the described experimental conditions, as the first one rapidly went rotten and the second one was suitable only on the day of harvest. There is evidence that these two species release ethylene [109,110], the hormone that affects the growth, development, and storage of many vegetables, fruits, and ornamental plants [111].

Despite the presence of ethylene can enhance coloration, it can also induce yellowing of green portions and softening, fostering the senescence of the stored material even at extremely low concentrations (30 ppb). The presence and dynamics of this plant growth regulator should be further investigated to understand its role during postharvest storage of *C. intybus* and *T. officinale* and generally all the edible flowers. During senescence, polyphenols are also possibly involved in affecting the visual quality of vegetable products. Most polyphenols are located in the vacuole of plant cells and once a physical stress or deteriorative process start, plant cells begin to break. Therefore, phenolic compounds mix with phenol peroxidases or polyphenol oxidase present in the cytoplasm and other cell organs, leading to the appearance of browning tissues [112].

Results indicated that the studied edible flowers maintained an acceptable visual quality for one week under cold storage while roses could last more. Further studies are thus necessary to explore the storage requirements of fresh edible flowers to maintain their good visual quality for longer periods and prevent flower damages (i.e., tissue browning, petal discoloration, or dehydration) [37,46].

4.3. Bioactive Compounds and Antioxidant Activity Dynamics

In this study, *P. officinalis*, *R. canina*, and *R. pendulina* had the highest values in polyphenols and antioxidant activity at harvest, confirming previous results on the same species [13,22,113–115]. Comparing the total phenolic compounds with previous studies, the range of values detected (76.41–1270.72 mg GAE/100 g FW) is in accordance with that found in 51 Chinese edible flowers [22] and in the methanolic extracts of five species [116]. Focusing on single species, *B. officinalis* polyphenols were similar to another research [15], but three-fold higher than another study [117]. Polyphenols in *C. cyanus* were not abundant, and 2.5-fold lower than the research of [9]. Results on *P. officinalis* were instead similar to that of the tree peony (*Paeonia*, section *Moutan*) cultivars [114]. Data of *S. nigra* polyphenols were three-fold lower than previously evaluated [118] and finally, the phenolic content of *T. majus* and *P. veris* was concordant with other reports on the same species [5,59]. Regarding anthocyanins, *S. pratensis* and *T. majus* showed the highest content, probably thanks to their bright blue and red-orange colors, determined by these phytochemicals [119]. However, comparisons with other studies are difficult for the lack of information on the 17 studied species.

Most of the 17 species had FRAP values similar to Chinese edible flowers [22], except for *P. officinalis*, both roses and *T. majus*, which showed higher levels of antioxidant activity with this assay. *Cichorium intybus* had threefold higher FRAP activity than the ethanolic extracts of the same species [18], while for *S. nigra* flowers similar values were reported [18]. The ranges of the radical scavenging assays (DPPH and ABTS) were also comparable to literature, except again for *P. officinalis* and roses which had higher values in this study. Conversely, *C. officinalis* FRAP and DPPH values and *T. majus* DPPH and ABTS values resulted lower than previous studies [64,120]. At harvest, our data support previous reports [17,23,115,116,121], indicating that the antioxidant activity is highly correlated with the phenolic pattern, as the polyphenols are among the main phytochemicals responsible for the antioxidant activity [7,23].

Polyphenol content in plant organs depends on several preharvest factors, mainly related to environmental and stress conditions, since they are principally produced as a defense against pathogens and solar radiation [122]. Some polyphenols are present in all plant products, while others are specific to particular foods, but mostly, plants have complex mixtures of phenolic compounds, which are often poorly characterized and can behave differently during storage and ripening. Few trials have been performed so far on the dynamics of bioactive compounds and antioxidant activity in edible flowers during storage, as the visual quality has been the most frequent evaluated parameter [11]. According to some authors [21], cold storage could affect the bioactive compounds of edible flowers. A reduction in total phenolic content in squash (*Cucurbita pepo* L.) flowers during storage at 5 °C was found [44], while no variations were detected in *A. oleracea* and *B. semperflorens* flowers stored at 4 °C [21,123]. Total polyphenol content and DPPH clearance activity changed slight in daylily flowers (*Hemerocallis lilioasphodelus* L.) during four days at 4 °C [43]. However, during storage throughout 20 days at 4 °C, the content of bioactive compounds and the antioxidant activity measured with DPPH assay increased in pansies [107]. In addition to differences among species, a comparison between edible flowers in plastic boxes and flow packs (stored at 4 °C) showed that the phytochemical content can also vary according to the packaging during postharvest [45].

In this study, the content of bioactive compounds differed considering the two-week trial, as polyphenols followed a different trend according to the species. Anthocyanins were less variable during storage, with no changes in 15 of the examined species, suggesting that anthocyanins were not influenced by cold storage. Among the antioxidant assays, ABTS values were the least variable, varying in only four species across the trial. Interestingly, *S. pratensis* was the only species where no variations in the five evaluated parameters occurred during the trial. *Tropaeolum majus* phenolic content and *S. pratensis* anthocyanin content were stable during storage, conversely to previous studies [21,108]. In addition, we recorded a decrease in anthocyanins and in antioxidant activity (especially for the FRAP assay) of nasturtium, conversely to previous findings [21].

The limit of visual acceptance was recorded at day 7 for most of the species (day 10 in *R. canina* and day 14 in *R. pendulina*). During the shelf life period, the total polyphenols slightly decreased in *C. officinalis* (−21.3%), *P. officinalis* (−4.9%), and *R. pendulina* (−12.5%), while increased to a higher extent in *C. cyanus* (+40.2%), *D. carthusianorum* (+43.7%), *L. vulgare* (+46.7%), *P. veris* (+26.4%), and *R. canina* (+30.6%). Generally, it has been seen that during ripening the concentration of phenolic acids decrease, while the anthocyanins increase. However, many factors can affect the content of polyphenols in plants, and different behavior have been recorded according to the species [124]. This can possibly be explained by the wide diversity of phenolic compounds that can be synthetized by plants, leading to different phenolic profiles [122]. In addition, if we consider that the dehydration of fresh material could occur during storage, affecting the amounts of bioactive compounds and antioxidant activity detected. Besides, an increased accumulation of phenolic compounds has been found to be related to exposure to ethylene in lettuce, asparagus, and parsnip during storage [111]. It is not currently known whether *C. cyanus*, *D. carthusianorum*, *L. vulgare*, *P. veris*, and *R. canina* produce ethylene except for the related species of carnation (*Dianthus caryophyllus*) and rose (*Rosa bourboniana* and *Rosa hybrida*), which are ethylene-sensitive [125–127]. Further investigations on this hormone production by edible flowers are thus needed to understand if and how it can affect phenolic dynamics during storage.

Polyphenols have also been found to decline (−46%) during the shelf life period of *B. semperflorens* (nine days), but was unaltered in *V. cornuta* (16 days) [45]. As per the antioxidant activity, the trend varied depending on the analytical method used; however, three species (*A. ursinum*, *C. cyanus*, and *R. canina*) showed increased values in all assays, DPPH (+99.6%, +59.1%, +34.3%), ABTS (+23%, +64.9%, +32.1%), and FRAP (+247.6%, +90.1%, +30.1%), compared to the day of harvest. A previous study [45] reported decreasing values in FRAP antioxidant activity during the flowers' shelf life (−52% in *B. semperflorens* and −34% in *V. cornuta*).

Numerous structural, physiological, and biochemical changes occur during ripening of horticultural products, which is a complex process [128]. This process is influenced by endogenous and environmental factors, involving multiple transcription regulatory and biochemical pathways that are still need to be clarified [128]. So far, cold storage has been seen to successfully delay flower senescence and quality deterioration of edible flowers, by slowing the growth of microorganisms and the production of ethylene, and by reducing internal breakdown of tissues, respiration, water loss and wilting [11]. To fully understand flower senescence during postharvest, more information on the complex phytochemical profile of each species are needed. However, data of dynamics during postharvest suggest that the studied edible flowers can be valuable sources of bioactive compounds exerting antioxidant activity, also beyond the limit of acceptance for sale purposes. Decaying fresh flowers can be thus recovered and not wasted and can be proposed for the extraction of valuable bioactive compounds.

5. Conclusions

This study described for the first time the sensory profiles of several edible flower species. The methodology presented here might be useful for the selection of sensory descriptors and for giving an indication on the range intensity values concerning the flowers, even if data will have to be further confirmed in future studies. All the species were also investigated for their main phytochemical characteristics related to bioactive compounds, showing a wide variability between species. Generally, cold storage (4 °C) seemed not to have negative effects on the phytochemical compounds, as the total phenolic and anthocyanin contents remained almost unaltered or even increased across 14 days. *Paeonia officinalis* exhibited the highest values in four out of five characteristics (total polyphenols, DPPH, ABTS, FRAP). Nevertheless, *P. officinalis* is not currently cultivated for its flowers and in North-West Italy (where the flowers were sampled) is a protected species.

Similar interesting results have been obtained in commonly cultivated roses. *Rosa canina* and *R. pendulina* flowers had 10 and 14 days of shelf life at 4 °C in plastic boxes, respectively, being interesting products to be sold as edible flowers, with *R. canina* having the strongest smell and rose aroma intensity. The high polyphenol content of these species might be responsible for their bitter taste, which must be considered before consumption or preparation of foods. *Salvia pratensis* too could be easily marketable, as its values did not change in all five parameters assessed during seven days of storage and it is characterized by high smell intensity and easy chewability. Edible flowers can be stored satisfactorily at 4 °C for 7–14 days according to the species, and used to confer appeal and a wide range of tastes, aromas and sensory characteristics to dishes or food products. The preliminary results on the subjective evaluation should be confirmed in the future by specific consumer's tests on a large number of individuals to provide indication of the possible final user's judgement. Despite its availability, it is eaten in lower quantities than fruits and vegetables. However, flowers are confirmed as source of bioactive compounds with antioxidant activity, which can provide not only aesthetic beauty but also benefits for health, explaining the increasing number of studies on new food applications of edible flowers.

Supplementary Materials: The following are available online at https://www.mdpi.com/article/10.3390/horticulturae7070166/s1, Figure S1: Evaluation sheet developed for the Quantitative Descriptive Analysis of edible flowers. Figure S2: Spider charts with intensities (from 0 to 10) of each descriptor detected in the flowers of *Allium ursinum*, *Borago officinalis*, *Calendula officinalis*, *Centaurea cyanus*, *Cichorium intybus*, and *Dianthus carthusianorum*. Figure S3: Spider charts with intensities (from 0 to 10) of each descriptor detected in the flowers of *Lavandula angustifolia*, *Leucanthemum vulgare*, *Paeonia officinalis*, *Primula veris*, *Robinia pseudoacacia*, and *Rosa canina*. Figure S4: Spider charts with intensities (from 0 to 10) of each descriptor detected in the flowers of *Rosa pendulina*, *Salvia pratensis*, *Sambucus nigra*, *Taraxacum officinale*, and *Tropaeolum majus*.

Author Contributions: Conceptualization, V.S.; methodology, M.G.M., V.S.; investigation, S.D., M.G.M., N.M.F., M.C.; resources, V.S.; data curation, S.D., M.G.M., N.M.F., M.C.; writing—original draft preparation, S.D., M.G.M., N.M.F.; writing—review and editing, S.D., M.G.M., N.M.F., M.C., V.S.; supervision, V.S.; funding acquisition, V.S. All authors have read and agreed to the published version of the manuscript.

Funding: This research was supported by the program Interreg V-A Francia Italia Alcotra, project n. 1139: "ANTEA—Attività innovative per lo sviluppo della filiera transfrontaliera del fiore edule".

Institutional Review Board Statement: Not applicable.

Informed Consent Statement: Not applicable.

Data Availability Statement: Data is contained within the article or Supplementary Material.

Acknowledgments: Regione Piemonte (Italy) is thanked for granting the sampling of flowers of the protected species *Paeonia officinalis* L. The authors thank Marie-Jeanne Welter and the Masters students of the Laboratory of Cultivation and Transformation of Aromatic Plants at the University of Turin for their help in collecting data during postharvest of edible flowers, and the Italian National Organization of Fruit's Tasters (O.N.A.Frut) for performing the sensory evaluations.

Conflicts of Interest: The authors declare no conflict of interest.

References

1. Mlcek, J.; Rop, O. Fresh edible flowers of ornamental plants—A new source of nutraceutical foods. *Trends Food Sci. Technol.* **2011**, *22*, 561–569. [CrossRef]
2. Fernandes, L.; Casal, S.; Pereira, J.A.; Saraiva, J.A.; Ramalhosa, E. Edible flowers: A review of the nutritional, antioxidant, antimicrobial properties and effects on human health. *J. Food Compos. Anal.* **2017**, *60*, 38–50. [CrossRef]
3. Scariot, V.; Gaino, W.; Demasi, S.; Caser, M.; Ruffoni, B. Flowers for edible gardens: Combinations of species and colours for northwestern Italy. *Acta Hortic.* **2018**, *1215*, 363–368. [CrossRef]
4. Lim, T.K. *Edible Medicinal and Non-Medicinal Plants. Volume 7, Flowers*; Springer: Dordrecht, The Netherlands, 2014, ISBN 978-94-007-7394-3.
5. Lim, T.K. *Edible Medicinal and Non-Medicinal Plants. Volume 8, Flowers*; Springer: Dordrecht, The Netherlands, 2014, ISBN 978-94-017-8747-5.
6. Chitrakar, B.; Zhang, M.; Bhandari, B. Edible flowers with the common name "marigold": Their therapeutic values and processing. *Trends Food Sci. Technol.* **2019**, *89*, 76–87. [CrossRef]
7. Koike, A.; Barreira, J.C.M.; Barros, L.; Santos-Buelga, C.; Villavicencio, A.L.C.H.; Ferreira, I.C.F.R. Edible flowers of *Viola tricolor* L. as a new functional food: Antioxidant activity, individual phenolics and effects of gamma and electron-beam irradiation. *Food Chem.* **2015**, *179*, 6–14. [CrossRef]
8. Chen, N.H.; Wei, S. Factors influencing consumers' attitudes towards the consumption of edible flowers. *Food Qual. Prefer.* **2017**, *56*, 93–100. [CrossRef]
9. Rop, O.; Mlcek, J.; Jurikova, T.; Neugebauerova, J.; Vabkova, J. Edible Flowers—A New Promising Source of Mineral Elements in Human Nutrition. *Molecules* **2012**, *17*, 6672–6683. [CrossRef]
10. Cunningham, E. What nutritional contribution do edible flowers make? *J. Acad. Nutr. Diet.* **2015**, *115*, 856. [CrossRef]
11. Fernandes, L.; Saraiva, J.A.; Pereira, J.A.; Casal, S.; Ramalhosa, E. Post-harvest technologies applied to edible flowers: A review. *Food Rev. Int.* **2019**, *35*, 132–154. [CrossRef]
12. Falla, N.M.; Contu, S.; Demasi, S.; Caser, M.; Scariot, V. Environmental impact of edible flower production: A case study. *Agronomy* **2020**, *10*, 579. [CrossRef]
13. Demasi, S.; Caser, M.; Donno, D.; Ravetto Enri, S.; Lonati, M.; Scariot, V. Exploring wild edible flowers as a source of bioactive compounds: New perspectives in horticulture. *Folia Hortic.* **2021**, *33*, 1–22. [CrossRef]
14. Pires, T.C.S.P.; Barros, L.; Santos-Buelga, C.; Ferreira, I.C.F.R. Edible flowers: Emerging components in the diet. *Trends Food Sci. Technol.* **2019**, *93*, 244–258. [CrossRef]

15. Grzeszczuk, M.; Stefaniak, A.; Pachlowska, A. Biological value of various edible flower species. *Sci. Pol.* **2016**, *15*, 109–119.
16. González-Barrio, R.; Periago, M.J.; Luna-Recio, C.; Garcia-Alonso, F.J.; Navarro-González, I. Chemical composition of the edible flowers, pansy (*Viola wittrockiana*) and snapdragon (*Antirrhinum majus*) as new sources of bioactive compounds. *Food Chem.* **2018**, *252*, 373–380. [CrossRef]
17. Kaisoon, O.; Konczak, I.; Siriamornpun, S. Potential health enhancing properties of edible flowers from Thailand. *Food Res. Int.* **2012**, *46*, 563–571. [CrossRef]
18. Loizzo, M.R.; Pugliese, A.; Bonesi, M.; Tenuta, M.C.; Menichini, F.; Xiao, J.; Tundis, R. Edible flowers: A rich source of phytochemicals with antioxidant and hypoglycemic properties. *J. Agric. Food Chem.* **2016**, *64*, 2467–2474. [CrossRef]
19. Ceccanti, C.; Landi, M.; Benvenuti, S.; Pardossi, A.; Guidi, L. Mediterranean wild edible plants: Weeds or "new functional crops". *Molecules* **2018**, *23*, 2299. [CrossRef]
20. Durazzo, A.; Lucarini, M.; Souto, E.B.; Cicala, C.; Caiazzo, E.; Izzo, A.A.; Novellino, E.; Santini, A. Polyphenols: A concise overview on the chemistry, occurrence, and human health. *Phyther. Res.* **2019**, *33*, 2221–2243. [CrossRef]
21. Landi, M.; Ruffoni, B.; Combournac, L.; Guidi, L. Nutraceutical value of edible flowers upon cold storage. *Ital. J. Food Sci.* **2018**, *30*, 336–347. [CrossRef]
22. Li, A.-N.; Li, S.; Li, H.-B.; Xu, D.-P.; Xu, X.-R.; Chen, F. Total phenolic contents and antioxidant capacities of 51 edible and wild flowers. *J. Funct. Foods* **2014**, *6*, 319–330. [CrossRef]
23. Lu, B.; Li, M.; Yin, R. Phytochemical content, health benefits, and toxicology of common edible flowers: A review (2000–2015). *Crit. Rev. Food Sci. Nutr.* **2016**, *56*, S130–S148. [CrossRef]
24. Zheng, J.; Yu, X.; Maninder, M.; Xu, B. Total phenolics and antioxidants profiles of commonly consumed edible flowers in China. *Int. J. Food Prop.* **2018**, *21*, 1524–1540. [CrossRef]
25. Fernandes, L.; Casal, S.; Pereira, J.A.; Saraiva, J.A.; Ramalhosa, E. An overview on the market of edible flowers. *Food Rev. Int.* **2020**, *36*, 258–275. [CrossRef]
26. Bursać, D.; Vahčić, N.; Levaj, B.; Dragović-Uzelac, V.; Biško, A. The influence of cultivar on sensory profiles of fresh and processed strawberry fruits grown in Croatia. *Flavour Fragr. J.* **2007**, *22*, 512–520. [CrossRef]
27. Rodrigues, H.; Cielo, D.P.; Goméz-Corona, C.; Silveira, A.A.S.; Marchesan, T.A.; Galmarini, M.V.; Richards, N.S.P.S. Eating flowers? Exploring attitudes and consumers' representation of edible flowers. *Food Res. Int.* **2017**, *100*, 227–234. [CrossRef]
28. Lawless, H.T.; Heymann, H. *Sensory Evaluation of Food-Principles and Practices*; Food Science Text Series; Springer: New York, NY, USA, 2010; Volume 24, ISBN 978-1-4419-6487-8.
29. Fernandes, L.; Casal, S.; Pereira, J.A.; Malheiro, R.; Rodrigues, N.; Saraiva, J.A.; Ramalhosa, E. Borage, calendula, cosmos, Johnny Jump up, and pansy flowers: Volatiles, bioactive compounds, and sensory perception. *Eur. Food Res. Technol.* **2019**, *245*, 593–606. [CrossRef]
30. Sharif, M.K.; Butt, M.S.; Sharif, H.R.; Nasir, M. Sensory Evaluation and Consumer Acceptability. In *Handbook of Food Science and Technology*; University of Agriculture: Faisalabad, Pakistan, 2017; pp. 362–386.
31. Benvenuti, S.; Bortolotti, E.; Maggini, R. Antioxidant power, anthocyanin content and organoleptic performance of edible flowers. *Sci. Hortic.* **2016**, *199*, 170–177. [CrossRef]
32. Kelley, K.M.; Behe, B.K.; Biernbaum, J.A.; Poff, K.L. Consumer preference for edible-flower color, container size, and price. *HortScience* **2001**, *36*, 801–804. [CrossRef]
33. Kelley, K.M.; Behe, B.K.; Biernbaum, J.A.; Poff, K.L. Consumer and professional chef perceptions of three edible-flower species. *HortScience* **2001**, *36*, 162–166. [CrossRef]
34. Kelley, K.M.; Behe, B.K.; Biernbaum, J.A.; Poff, K.L. Combinations of colors and species of containerized edible flowers: Effect on consumer preferences. *HortScience* **2002**, *37*, 218–221. [CrossRef]
35. Simoni, N.K.; Santos, F.F.; Andrade, T.A.; Villavicencio, A.L.C.H.; Pinto-e-Silva, M.E.M. The use of edible flowers in human food: Sensory analysis of preparations. *ETP Int. J. Food Eng.* **2018**, 140–143. [CrossRef]
36. D'Antuono, L.F.; Manco, M.A. Preliminary sensory evaluation of edible flowers from wild *Allium* species. *J. Sci. Food Agric.* **2013**, *93*, 3520–3523. [CrossRef] [PubMed]
37. Kou, L.; Turner, E.R.; Luo, Y. Extending the shelf life of edible flowers with controlled release of 1-Methylcyclopropene and modified atmosphere packaging. *J. Food Sci.* **2012**, *77*, S188–S193. [CrossRef] [PubMed]
38. Kelley, K.M.; Cameron, A.C.; Biernbaum, J.A.; Poff, K.L. Effect of storage temperature on the quality of edible flowers. *Postharvest Biol. Technol.* **2003**, *27*, 341–344. [CrossRef]
39. Flores-López, M.L.; Cerqueira, M.A.; de Rodríguez, D.J.; Vicente, A.A. Perspectives on utilization of edible coatings and nano-laminate coatings for extension of postharvest storage of fruits and vegetables. *Food Eng. Rev.* **2016**, *8*, 292–305. [CrossRef]
40. Edelman, N.F.; Jones, M.L. Evaluating ethylene sensitivity within the family Solanaceae at different developmental stages. *HortScience* **2014**, *49*, 628–636. [CrossRef]
41. Edelman, N.F.; Kaufman, B.A.; Jones, M.L. Comparative evaluation of seedling hypocotyl elongation and mature plant assays for determining ethylene sensitivity in bedding plants. *HortScience* **2014**, *49*, 472–480. [CrossRef]
42. Macnish, A.J.; Jiang, C.-Z.; Negre-Zakharov, F.; Reid, M.S. Physiological and molecular changes during opening and senescence of *Nicotiana mutabilis* flowers. *Plant Sci.* **2010**, *179*, 267–272. [CrossRef]
43. Liu, W.; Zhang, J.; Zhang, Q.; Shan, Y. Effects of postharvest chilling and heating treatments on the sensory quality and antioxidant system of daylily flowers. *Hortic. Environ. Biotechnol.* **2018**, *59*, 671–685. [CrossRef]

44. Aquino-Bolaños, E.N.; Urrutia-Hernández, T.A.; López Del Castillo-Lozano, M.; Chavéz-Servia, J.L.; Verdalet-Guzmán, I. Physicochemical parameters and antioxidant compounds in edible squash (*Cucurbita pepo*) flower stored under controlled atmospheres. *J. Food Qual.* **2013**, *36*, 302–308. [CrossRef]
45. Demasi, S.; Falla, N.M.; Caser, M.; Scariot, V. Postharvest aptitude of *Begonia semperflorens* and *Viola cornuta* edible flowers. *Adv. Hortic. Sci.* **2020**, *34*, 13–20. [CrossRef]
46. Chrysargyris, A.; Tzionis, A.; Xylia, P.; Tzortzakis, N. Effects of salinity on tagetes growth, physiology, and shelf life of edible flowers stored in passive modified atmosphere packaging or treated with ethanol. *Front. Plant Sci.* **2018**, *871*. [CrossRef]
47. Pavlović, D.R.; Veljković, M.; Stojanović, N.M.; Gočmanac-Ignjatović, M.; Mihailov-Krstev, T.; Branković, S.; Sokolović, D.; Marčetić, M.; Radulović, N.; Radenković, M. Influence of different wild-garlic (*Allium ursinum*) extracts on the gastrointestinal system: Spasmolytic, antimicrobial and antioxidant properties. *J. Pharm. Pharmacol.* **2017**, *69*, 1208–1218. [CrossRef]
48. Bombicz, M.; Priksz, D.; Varga, B.; Kurucz, A.; Kertész, A.; Takacs, A.; Posa, A.; Kiss, R.; Szilvassy, Z.; Juhasz, B. A novel therapeutic approach in the treatment of pulmonary arterial hypertension: *Allium ursinum* liophylisate alleviates symptoms comparably to Sildenafil. *Int. J. Mol. Sci.* **2017**, *18*, 1436. [CrossRef]
49. Sobolewska, D.; Podolak, I.; Makowska-Wąs, J. *Allium ursinum*: Botanical, phytochemical and pharmacological overview. *Phytochem. Rev.* **2015**, *14*, 81–97. [CrossRef]
50. Karimi, E.; Oskoueian, E.; Karimi, A.; Noura, R.; Ebrahimi, M. *Borago officinalis* L. flower: A comprehensive study on bioactive compounds and its health-promoting properties. *J. Food Meas. Charact.* **2018**, *12*, 826–838. [CrossRef]
51. Ukiya, M.; Akihisa, T.; Yasukawa, K.; Tokuda, H.; Suzuki, T.; Kimura, Y. Anti-inflammatory, anti-tumor-promoting, and cytotoxic activities of constituents of marigold (*Calendula officinalis*) flowers. *J. Nat. Prod.* **2006**, *69*, 1692–1696. [CrossRef]
52. Muley, B.P.; Khadabadi, S.S.; Banarase, N.B. Phytochemical constituents and pharmacological activities of *Calendula officinalis* L. (Asteraceae): A review. *Trop. J. Pharm. Res.* **2009**, *8*, 455–465. [CrossRef]
53. Palma, L. *Le Piante Medicinali d'Italia*; Società Editrice Internazionale: Torino, Italy, 1964.
54. Rashed, M.M.A.; Tong, Q.; Nagi, A.; Li, J.; Khan, N.U.; Chen, L.; Rotail, A.; Bakry, A.M. Isolation of essential oil from *Lavandula angustifolia* by using ultrasonic-microwave assisted method preceded by enzymolysis treatment, and assessment of its biological activities. *Ind. Crops Prod.* **2017**, *100*, 236–245. [CrossRef]
55. Facciola, S. *Cornucopia: A Source Book of Edible Plants*; Kampong Publications: Vista, CA, USA, 1990; ISBN 0-9628087-0-9.
56. Magharri, E.; Razavi, S.M.; Ghorbani, E.; Nahar, L.; Sarker, S.D. Chemical composition, some allelopathic aspects, free-radical-scavenging property and antifungal activity of the volatile oil of the flowering tops of *Leucanthemum vulgare* Lam. *Rec. Nat. Prod.* **2015**, *9*, 538–545.
57. Chen, Z.; Li, X.-P.; Li, Z.-J.; Xu, L.; Li, X.-M. Reduced hepatotoxicity by total glucosides of paeony in combination treatment with leflunomide and methotrexate for patients with active rheumatoid arthritis. *Int. Immunopharmacol.* **2013**, *15*, 474–477. [CrossRef] [PubMed]
58. Ghédira, K.; Goetz, P. Pivoine *Paeonia officinalis* L. (Paeoniaceae). *Phytothérapie* **2015**, *13*, 328–331. [CrossRef]
59. Tünde, J.; Eleonora, M.; Laura, V.; Neagu, O.; Annamária, P. Bioactive compounds and antioxidant capacity of *Primula veris* L. flower extracts. *Ecotoxicologie Zooteh. Ind.* **2015**, *15B*, 235–242.
60. Lattanzio, F.; Greco, E.; Carretta, D.; Cervellati, R.; Govoni, P.; Speroni, E. In vivo anti-inflammatory effect of *Rosa canina* L. extract. *J. Ethnopharmacol.* **2011**, *137*, 880–885. [CrossRef] [PubMed]
61. Guarrera, P.M.; Savo, V. Wild food plants used in traditional vegetable mixtures in Italy. *J. Ethnopharmacol.* **2016**, *185*, 202–234. [CrossRef]
62. Schütz, K.; Carle, R.; Schieber, A. Taraxacum—A review on its phytochemical and pharmacological profile. *J. Ethnopharmacol.* **2006**, *107*, 313–323. [CrossRef]
63. Martinez, M.; Poirrier, P.; Chamy, R.; Prüfer, D.; Schulze-Gronover, C.; Jorquera, L.; Ruiz, G. *Taraxacum officinale* and related species—An ethnopharmacological review and its potential as a commercial medicinal plant. *J. Ethnopharmacol.* **2015**, *169*, 244–262. [CrossRef]
64. Garzón, G.A.; Wrolstad, R.E. Major anthocyanins and antioxidant activity of Nasturtium flowers (*Tropaeolum majus*). *Food Chem.* **2009**, *114*, 44–49. [CrossRef]
65. Bazylko, A.; Granica, S.; Filipek, A.; Piwowarski, J.; Stefańska, J.; Osińska, E.; Kiss, A.K. Comparison of antioxidant, anti-inflammatory, antimicrobial activity and chemical composition of aqueous and hydroethanolic extracts of the herb of *Tropaeolum majus* L. *Ind. Crops Prod.* **2013**, *50*, 88–94. [CrossRef]
66. Canterino, S.; Donno, D.; Mellano, M.G.; Beccaro, G.L.; Bounous, G. Nutritional and sensory survey of *Citrus sinensis* (L.) cultivars grown at the most northern limit of the mediterranean latitude. *J. Food Qual.* **2012**, *35*, 108–118. [CrossRef]
67. Donno, D.; Beccaro, G.L.; Mellano, M.G.; Torello Marinoni, D.; Cerutti, A.K.; Canterino, S.; Bounous, G. Application of sensory, nutraceutical and genetic techniques to create a quality profile of ancient apple cultivars. *J. Food Qual.* **2012**, *35*, 169–181. [CrossRef]
68. Kemp, S.E.; Hollowood, T.; Hort, J. *Sensory Evaluation: A Practical Handbook*; John Wiley & Sons, Ltd.: Chichester, UK, 2009, ISBN 978-1-405-16210-4.
69. Meilgaard, M.; Civille, G.V.; Carr, B.T. *Sensory Evaluation Techniques*; CRC Press: Boca Raton, FL, USA, 1999; ISBN 9780849302763.
70. Stone, H.; Sidel, J.; Oliver, S.; Woolsey, A.; Singleton, R.C. Sensory evaluation by Quantitative Descriptive Analysis. In *Descriptive Sensory Analysis in Practice*; Food & Nutrition Press, Inc.: Trumbull, CT, USA, 2008; pp. 23–34.

71. De Biaggi, M.; Rapalino, S.; Donno, D.; Mellano, M.G.; Beccaro, G.L. Genotype influence on chemical composition and sensory traits of chestnut in 18 cultivars grown on the same rootstock and at the same agronomic conditions. *Acta Hortic.* **2018**, 215–220. [CrossRef]
72. Mellano, M.G.; Carli, C.; Folini, L.; Draicchio, P.; Beccaro, G. Training of two groups of tasters for the creation of sensory profiles of highbush blueberry cultivars grown in Northern Italy. *Acta Hortic.* **2009**, *810*, 835–840. [CrossRef]
73. Mellano, M.G.; Rapalino, S.; Donno, D. Profili sensoriali di cultivar di *Castanea sativa* e ibridi eurogiapponesi (in Italian). *Castanea* **2017**, *9*, 8–9.
74. Montevecchi, G.; Mellano, M.G.; Simone, G.V.; Masino, F.; Bignami, C.; Antonelli, A. Physico-chemical and sensory characterization of pescabivona P.G.I., A sicilian white flesh peach cultivar [*Prunus persica* (L.) batsch]: A case study. In *Apricots and Peaches: Nutritional Properties, Post-Harvest Management and Potential Health Benefits*; Nova Science Publishers, Inc.: New York, NY, USA, 2016; pp. 93–124. ISBN 9781634846851.
75. Tobin, R.; Moane, S.; Larkin, T. Sensory evaluation of organic and conventional fruits and vegetables available to Irish consumers. *Int. J. Food Sci. Technol.* **2013**, *48*, 157–162. [CrossRef]
76. Zhao, X.; Chambers, E., IV; Matta, Z.; Loughin, T.M.; Carey, E.E. Consumer sensory analysis of organically and conventionally grown vegetables. *J. Food Sci.* **2007**, *72*, S87–S91. [CrossRef]
77. Caser, M.; Demasi, S.; Stelluti, S.; Donno, D.; Scariot, V. *Crocus sativus* L. cultivation in alpine environments: Stigmas and tepals as source of bioactive compounds. *Agronomy* **2020**, *10*, 1473. [CrossRef]
78. Singleton, V.L.; Orthofer, R.; Lamuela-Raventós, R.M. Analysis of total phenols and other oxidation substrates and antioxidants by means of folin-ciocalteu reagent. *Methods Enzymol.* **1999**, *299*, 152–178. [CrossRef]
79. Caser, M.; Victorino, Í.M.M.; Demasi, S.; Berruti, A.; Donno, D.; Lumini, E.; Bianciotto, V.; Scariot, V. Saffron cultivation in marginal alpine environments: How AMF inoculation modulates yield and bioactive compounds. *Agronomy* **2019**, *9*, 12. [CrossRef]
80. Lee, J.; Durst, R.W.; Wrolstad, R.E.; Eisele, T.; Giusti, M.M.; Hach, J.; Hofsommer, H.; Koswig, S.; Krueger, D.A.; Kupina, S.; et al. Determination of total monomeric anthocyanin pigment content of fruit juices, beverages, natural colorants, and wines by the pH differential method: Collaborative study. *J. AOAC Int.* **2005**, *88*, 1269–1278. [CrossRef]
81. Wong, S.P.; Leong, L.P.; William Koh, J.H. Antioxidant activities of aqueous extracts of selected plants. *Food Chem.* **2006**, *99*, 775–783. [CrossRef]
82. Benzie, I.F.F.; Strain, J.J. Ferric reducing/antioxidant power assay: Direct measure of total antioxidant activity of biological fluids and modified version for simultaneous measurement of total antioxidant power and ascorbic acid concentration. *Methods Enzymol.* **1998**, *299*, 15–27. [CrossRef]
83. Hammer, Ø.; Harper, D.A.T.; Ryan, P.D. PAST: Paleontological statistics software package for education and data analysis. *Palaeontol. Electron.* **2001**, *4*, 9.
84. Bachmanov, A.A. Genetic Architecture of Sweet Taste. In *ACS Symposium Series*; American Chemical Society: Washington, DC, USA, 2008; Volume 979, pp. 18–47. ISBN 978-0-84127-432-7.
85. Beauchamp, G.K. Basic taste: A perceptual concept. *J. Agric. Food Chem.* **2019**, *67*, 13860–13869. [CrossRef] [PubMed]
86. Shallenberger, R.S. Sugar structure and taste. In *Advances in Chemistry*; ACS Publications: Washington, DC, USA, 1973; pp. 256–263 ISBN 978-0-84120-178-1.
87. Zarzo, M. A sensory 3D map of the odor description space derived from a comparison of numeric odor profile databases. *Chem Senses* **2015**, *40*, 305–313. [CrossRef] [PubMed]
88. Forde, C.G.; van Kuijk, N.; Thaler, T.; de Graaf, C.; Martin, N. Oral processing characteristics of solid savoury meal components, and relationship with food composition, sensory attributes and expected satiation. *Appetite* **2013**, *60*, 208–219. [CrossRef]
89. Drewnowski, A. The science and complexity of bitter taste. *Nutr. Rev.* **2009**, *59*, 163–169. [CrossRef]
90. Simat, T. Panel Training on Odour and Aroma Perception for Sensory Analysis. Available online: https://www.dlg.org/en/food/topics/dlg-expert-reports/sensory-technology/dlg-expert-report-1-2017/ (accessed on 27 October 2017).
91. Vieira, P.M. Avaliação da Composição Química, dos Compostos Bioativos e da Atividade Antioxidante Em Seis Espécies de Flores Comestíveis, Universidade Estadual Paulista "Júlio De Mesquita Filho". 2013. Available online: http://hdl.handle.net/11449/100866 (accessed on 27 October 2017).
92. Byrnes, N.K. *The Influence of Experience and Personality on the Perception, Liking, and Intake of Spicy Foods*; Pennsylvania State University: State College, PA, USA, 2014.
93. Yang, P.; Song, H.; Wang, L.; Jing, H. Characterization of key Aroma-active compounds in black garlic by sensory-directed flavor analysis. *J. Agric. Food Chem.* **2019**, *67*, 7926–7934. [CrossRef]
94. Ashok, P.K.; Upadhyaya, K. Tannins are astringent. *J. Pharmacogn. Phytochem.* **2012**, *1*, 45–50.
95. Noble, A.C. Astringency and Bitterness of Flavonoid Phenols. In *Chemistry of Taste*; ACS Symposium Series; ACS Publications: Washington, DC, USA, 2002; Volume 825, pp. 192–201.
96. Ma, W.; Guo, A.; Zhang, Y.; Wang, H.; Liu, Y.; Li, H. A review on astringency and bitterness perception of tannins in wine. *Trends Food Sci. Technol.* **2014**, *40*, 6–19. [CrossRef]
97. Dinnella, C.; Recchia, A.; Tuorila, H.; Monteleone, E. Individual astringency responsiveness affects the acceptance of phenol-rich foods. *Appetite* **2011**, *56*, 633–642. [CrossRef]
98. Neta, E.R.D.C.; Johanningsmeier, S.D.; McFeeters, R.F. The chemistry and physiology of sour taste—A review. *J. Food Sci.* **2007**, *72*, R33–R38. [CrossRef]

99. Rocha, I.F.D.O.; Bolini, H.M.A. Passion fruit juice with different sweeteners: Sensory profile by descriptive analysis and acceptance. *Food Sci. Nutr.* **2015**, *3*, 129–139. [CrossRef]
100. Melgarejo, P.; Calín-Sánchez, Á.; Carbonell-Barrachina, Á.A.; Martínez-Nicolás, J.J.; Legua, P.; Martínez, R.; Hernández, F. Antioxidant activity, volatile composition and sensory profile of four new very-early apricots (*Prunus armeniaca* L.). *J. Sci. Food Agric.* **2014**, *94*, 85–94. [CrossRef]
101. Hernández, F.; Noguera-Artiaga, L.; Burló, F.; Wojdyło, A.; Carbonell-Barrachina, Á.A.; Legua, P. Physico-chemical, nutritional, and volatile composition and sensory profile of Spanish jujube (*Ziziphus jujuba* Mill.) fruits. *J. Sci. Food Agric.* **2016**, *96*, 2682–2691. [CrossRef]
102. Kirsanov, D.; Mednova, O.; Vietoris, V.; Kilmartin, P.A.; Legin, A. Towards reliable estimation of an "electronic tongue" predictive ability from PLS regression models in wine analysis. *Talanta* **2012**, *90*, 109–116. [CrossRef]
103. Cheong, M.W.; Liu, S.Q.; Zhou, W.; Curran, P.; Yu, B. Chemical composition and sensory profile of pomelo (*Citrus grandis* (L.) Osbeck) juice. *Food Chem.* **2012**, *135*, 2505–2513. [CrossRef]
104. Pinto, T.; Vilela, A.; Pinto, A.; Nunes, F.M.; Cosme, F.; Anjos, R. Influence of cultivar and of conventional and organic agricultural practices on phenolic and sensory profile of blackberries (*Rubus fruticosus*). *J. Sci. Food Agric.* **2018**, *98*, 4616–4624. [CrossRef]
105. Najar, B.; Demasi, S.; Caser, M.; Gaino, W.; Cioni, P.L.; Pistelli, L.; Scariot, V. Cultivation substrate composition influences morphology, volatilome and essential oil of *Lavandula angustifolia* Mill. *Agronomy* **2019**, *9*, 411. [CrossRef]
106. Caser, M.; Demasi, S.; Victorino, Í.M.M.; Donno, D.; Faccio, A.; Lumini, E.; Bianciotto, V.; Scariot, V. Arbuscular mycorrhizal fungi modulate the crop performance and metabolic profile of saffron in soilless cultivation. *Agronomy* **2019**, *9*, 232. [CrossRef]
107. Fernandes, L.; Casal, S.; Pereira, J.A.; Pereira, E.L.; Ramalhosa, E.; Saraiva, J.A. Effect of high hydrostatic pressure on the quality of four edible flowers: *Viola × wittrockiana, Centaurea cyanus, Borago officinalis* and *Camellia japonica*. *Int. J. Food Sci. Technol.* **2017**, *52*, 2455–2462. [CrossRef]
108. Landi, M.; Ruffoni, B.; Salvi, D.; Savona, M.; Guidi, L. Cold storage does not affect ascorbic acid and polyphenolic content of edible flowers of a new hybrid of sage. *Agrochimica* **2015**, *59*, 348–357. [CrossRef]
109. Stewart-Wade, S.M.; Neumann, S.; Collins, L.L.; Boland, G.J. The biology of Canadian weeds. 117 *Taraxacum officinale* G. H. Weber ex Wiggers. *Can. J. Plant Sci.* **2002**, *82*, 825–853. [CrossRef]
110. Sadeghi, H.; Ghanaatiyan, K. Probing the responses of four chicory ecotypes by ethylene accumulation and growth characteristics under drought stress. *Ital. J. Agron.* **2017**, *12*, 177–182. [CrossRef]
111. Saltveit, M.E. Effect of ethylene on quality of fresh fruits and vegetables. *Postharvest Biol. Technol.* **1999**, *15*, 279–292. [CrossRef]
112. Toivonen, P.M.A.; Brummell, D.A. Biochemical bases of appearance and texture changes in fresh-cut fruit and vegetables. *Postharvest Biol. Technol.* **2008**, *48*, 1–14. [CrossRef]
113. Kumar, N.; Bhandari, P.; Singh, B.; Bari, S.S. Antioxidant activity and ultra-performance LC-electrospray ionization-quadrupole time-of-flight mass spectrometry for phenolics-based fingerprinting of Rose species: *Rosa damascena, Rosa bourboniana* and *Rosa brunonii*. *Food Chem. Toxicol.* **2009**, *47*, 361–367. [CrossRef]
114. Fan, J.; Zhu, W.; Kang, H.; Ma, H.; Tao, G. Flavonoid constituents and antioxidant capacity in flowers of different Zhongyuan tree penoy cultivars. *J. Funct. Foods* **2012**, *4*, 147–157. [CrossRef]
115. Xiong, L.; Yang, J.; Jiang, Y.; Lu, B.; Hu, Y.; Zhou, F.; Mao, S.; Shen, C. Phenolic compounds and antioxidant capacities of 10 common edible flowers from China. *J. Food Sci.* **2014**, *79*, C517–C525. [CrossRef]
116. Petrova, I.; Petkova, N.; Ivanov, I. Five edible flowers—Valuable source of antioxidants in human nutrition. *Int. J. Pharmacogn. Phytochem. Res.* **2016**, *8*, 604–610.
117. Aliakbarlu, J.; Tajik, H. Antioxidant and antibacterial activities of various extracts of *Borago officinalis* flowers. *J. Food Process. Preserv.* **2012**, *36*, 539–544. [CrossRef]
118. Młynarczyk, K.; Walkowiak-Tomczak, D.; Łysiak, G.P. Bioactive properties of *Sambucus nigra* L. as a functional ingredient for food and pharmaceutical industry. *J. Funct. Foods* **2018**, *40*, 377–390. [CrossRef]
119. Khoo, H.E.; Azlan, A.; Tang, S.T.; Lim, S.M. Anthocyanidins and anthocyanins: Colored pigments as food, pharmaceutical ingredients, and the potential health benefits. *Food Nutr. Res.* **2017**, *61*, 1361779. [CrossRef]
120. Butnariu, M.; Coradini, C.Z. Evaluation of biologically active compounds from *Calendula officinalis* flowers using spectrophotometry. *Chem. Cent. J.* **2012**, *6*, 35. [CrossRef]
121. Ji, H.F.; Du, A.L.; Zhang, L.W.; Xu, C.Y.; Yang, M.D.; Li, F.F. Effects of drying methods on antioxidant properties in *Robinia pseudoacacia* L. flowers. *J. Med. Plants Res.* **2012**, *6*. [CrossRef]
122. Manach, C.; Scalbert, A.; Morand, C.; Rémésy, C.; Jiménez, L. Polyphenols: Food sources and bioavailability. *Am. J. Clin. Nutr.* **2004**, *79*, 727–747. [CrossRef]
123. Friedman, H.; Rot, I.; Agami, O.; Vinokur, Y.; Rodov, V.; Reznick, N.; Umiel, N.; Dori, I.; Ganot, L.; Shmuel, D.; et al. Edible flowers: New crops with potential health benefits. *Acta Hortic.* **2007**, *755*, 283–290. [CrossRef]
124. D'Archivio, M.; Filesi, C.; Varì, R.; Scazzocchio, B.; Masella, R. Bioavailability of the Polyphenols: Status and Controversies. *Int. J. Mol. Sci.* **2010**, *11*, 1321–1342. [CrossRef]
125. Pun, U.K.; Yamada, T.; Tanase, K.; Shimizu-Yumoto, H.; Satoh, S.; Ichimura, K. Effect of ethanol on ethylene biosynthesis and sensitivity in cut carnation flowers. *Postharvest Biol. Technol.* **2014**, *98*, 30–33. [CrossRef]
126. Khatami, F.; Najafi, F.; Yari, F.; Khavari-Nejad, R.A. Expression of etr1-1 gene in transgenic *Rosa hybrida* L. increased postharvest longevity through reduced ethylene biosynthesis and perception. *Sci. Hortic.* **2020**, *263*, 109103. [CrossRef]

127. Singh, P.; Bharti, N.; Singh, A.P.; Tripathi, S.K.; Pandey, S.P.; Chauhan, A.S.; Kulkarni, A.; Sane, A.P. Petal abscission in fragrant roses is associated with large scale differential regulation of the abscission zone transcriptome. *Sci. Rep.* **2020**, *10*, 17196. [CrossRef]
128. Gong, T.; Li, C.; Bian, B.; Wu, Y.; Dawuda, M.M.; Liao, W. Advances in application of small molecule compounds for extending the shelf life of perishable horticultural products: A review. *Sci. Hortic.* **2018**, *230*, 25–34. [CrossRef]

 horticulturae

Article

Phytochemical Profile and Antioxidant Properties of Italian Green Tea, a New High Quality Niche Product

Nicole Mélanie Falla, Sonia Demasi, Matteo Caser * and Valentina Scariot

Department of Agricultural, Forest and Food Sciences, University of Torino, Largo Paolo Braccini 2, 10095 Grugliasco, TO, Italy; nicolemelanie.falla@unito.it (N.M.F.); sonia.demasi@unito.it (S.D.); valentina.scariot@unito.it (V.S.)

* Correspondence: matteo.caser@unito.it; Tel.: +39-011-670-8935

Abstract: The hot beverage commonly known as tea results from the infusion of dried leaves of the plant *Camellia sinensis* (L.) O. Kuntze. Ranking second only to water for its consumption worldwide, it has always been appreciated since antiquity for its aroma, taste characteristics, and beneficial effects on human health. There are many different processed tea types, including green tea, a non-fermented tea which, due to oxidation prevention maintains the structure of the bioactive compounds, especially polyphenols; these bioactive compounds show a number of benefits for the human health. The main producers of tea are China and India, followed by Kenya, Sri Lanka, Turkey, and Vietnam, however recently new countries are entering the market, with quality niche productions, among which also Italy. The present research aimed to assess the bioactive compounds (polyphenols) and the antioxidant activity of two green teas (the "Camellia d'Oro" tea—TCO, and the "Compagnia del Lago" tea—TCL) produced in Italy, in the Lake Maggiore district, where nurserymen have recently started to cultivate *C. sinensis*. In this area the cultivation of acidophilic plants as ornamentals has been known since around 1820. Due to the crisis of the floricultural sector, producers have been trying to diversify their product in order to increase their competitiveness, starting to cultivate Italian tea. Their antioxidant activity was assessed, finding a similar or higher antioxidant capacity than in other green teas, as reported in literature. TCO showed a higher antioxidant activity (42,758.86 mmol Fe^{2+} kg^{-1}; 532.37 μmol TE g^{-1} DW; 881.08 μmol TE g^{-1} DW) and phenolic content (14,918.91 mg GAE 100 g^{-1} DW) than TCL (25,796.61 mmol Fe^{2+} kg^{-1}; 302.35 μmol TE g^{-1} DW; 623.44 μmol TE g^{-1} DW; 8540.42 mg GAE 100 g^{-1} DW). Through HPLC, a total of thirteen phenolic compounds were identified quantitatively, including catechins, benzoic acids, cinnamic acids, and flavonols, in TCO while only 9 in TCL, and mainly in lower amounts. Albeit with differences, both teas were found to be of quality proving that Italy could have the possibility to grow profitably *C. sinensis*.

Keywords: tea; *Camellia sinensis*; antioxidant activity; DPPH; ABTS; FRAP; HPLC

Citation: Falla, N.M.; Demasi, S.; Caser, M.; Scariot, V. Phytochemical Profile and Antioxidant Properties of Italian Green Tea, a New High Quality Niche Product. *Horticulturae* **2021**, *7*, 91. https://doi.org/10.3390/horticulturae7050091

Academic Editors: Lucia Guidi, Luigi De Bellis, Alberto Pardossi and Elazar Fallik

Received: 15 March 2021
Accepted: 23 April 2021
Published: 27 April 2021

1. Introduction

The hot beverage resulting from the infusion of dried leaves of the plant *Camellia sinensis* (L.) O. Kuntze is commonly known as tea [1,2]. Tea is one of the most popular nonalcoholic beverages consumed across the world, second only to water [3–9], so much appreciated since antiquity for its aroma, taste characteristics, and beneficial health effects [1], thus consumed as an herbal infuse and for its medicinal properties [5,6]. World tea consumption increased to 5.5 million tons by 2016, mostly due to a rapid growth in per capita income levels in China, India, and other emerging economies [10]. The World Bank foresees an increase in average tea auction price from USD 2.80 in 2017 to USD 2.84 in 2020, expecting an extension of the global tea market [11]. World tea production reached 5.73 million tons in 2016 [10], with China constituting of 42.6% of world tea production [10], accounting for USD 4 billion [12]. India, the second largest producer, recorded a production of 1.27 million tons in 2016.

There are many different types of teas, and most of them are prepared with the buds of two botanical varieties, *C. sinensis* var. *sinensis* and *C. sinensis* var. *assamica* (Masters) Wight [9,13]; their characteristics (i.e., appearance, organoleptic taste, chemical contents, and flavor) vary according to the fermentation level: tea can be categorized in nonfermented tea (i.e., green tea, and white tea), semifermented tea (i.e., oolong tea), and fermented tea (i.e., black tea and red tea) [5,6,9,14–17]. The 78% of the tea worldwide production is black tea, especially consumed in Western countries, 20% consists of green tea, which is usually consumed in Asian countries, and 2% regards Oolong tea, commonly consumed in southern China [14]. Globally, the production of green tea increased annually by 5.4% over the past decade, also due to green tea's perceived health benefits [10].

Actually, fresh tea leaves contain chemical components such as polyphenols (catechins, flavonoids), alkaloids (caffeine, theobromine, etc.), volatile oils, polysaccharides, amino acids, lipids, vitamins (e.g., vitamin C), etc. [6,9,18]. Green teas are produced through steaming or roasting, thus inactivating the activity of polyphenol oxidase, preventing oxidation and so maintaining the structure of the phenolic compounds [1,5,18]. The resulting green teas show a polyphenol content varying from 30% to 42% of dry matter weight [14].

Due to its chemical constituents, tea shows many beneficial properties, such as antioxidant, anti-inflammatory, antiallergic, anticarcinogenic, antidiabetic, and antimicrobial effects [6,7,9,14,15,19–21]. A regular, daily consumption of green tea has been associated with many health benefits [7,22], which are mainly attributed to polyphenols, especially catechins [3,9,23].

Khan and Mukhtar (2007) reported that a balanced diet and the consumption of green tea can protect from oxidative stress and reduce reactive oxygen species damages to lipid membranes, proteins, and nucleic acids.

Moreover, many epidemiological studies investigated how tea consumption affected the incidence of cancer in humans, finding a protective and preventive effect of tea against various types of cancer (oral, pharyngeal, and laryngeal cancer) [14], but also healthy effects on many other pathologies involving oxidative stress.

Today, more than 50 countries produce different types of tea worldwide, not only as an herbal infuse pleasant to consume, but also for its well-known benefits on human health [2,5].

Tea evergreen plant (*C. sinensis*) [3,4] is native to South and Southwest China, the Indian Subcontinent, and Southeast Asia [2,24,25], then it became popular in India and Japan, and later in Europe and Russia [5,6]. The first plant specimens arrived in Europe were those studied by Linnaeus in 1763 [26], although the tea beverage had already reached Europe in the XVII century. However, tea cultivation has remained a prerogative of Asian countries, while in Europe the cultivation of the congeneric species *Camellia japonica* has spread for ornamental purposes only. The first camellia (*C. japonica*) was introduced in Italy at the end of the XVII century, in Caserta (Campania, South Italy) [27,28]. Since then its cultivation gradually became popular, reaching central and northern Italy, especially Tuscany, Piedmont, and Lombardy [27].

The critical period of the floricultural sector caused by globalization [27,29], since the end of the past decade forced Italian producers of the Lake Maggiore district (Piedmont region) to diversify their final products in order to increase their competitiveness, starting to cultivate *C. sinensis* in order to produce Italian green tea.

Currently there are no studies related to the quality of tea produced in Italy, thus, the aim of the present research was to assess the main bioactive compounds (polyphenols), and beneficial properties (antioxidant activity) of two green teas produced in this new productive context.

2. Materials and Methods

2.1. Plant Material and Site Characteristics

Camellia sinensis var. *sinensis* dried leaves were kindly provided by "La Compagnia del Lago" and "La Camelia d'Oro" plantations, both located in the Lake Maggiore area, in Piedmont—North Italy. Seedlings of camellia derived from acclimatization specimen from parks and botanical gardens located in the Lake Maggiore area (i.e., Villa Taranto, Isola Madre and Villa Anelli, Verbania municipality, Piedmont region), and were grown differently in the two nurseries. The "Compagnia del Lago" plantation is located in the municipality of Premosello Chiovenda (VB) (45°55′57.6″ N 8°27′16.1″ E), where the annual average maximum temperature was 19.8 °C (July showed the highest monthly average temperature, with 30.6 °C), the annual average minimum temperature was 7.8 °C (January showed the lowest monthly average temperature, with −3.6 °C), and the annual average rainfall was 122.8 mm (March showed the highest monthly average rainfall, with 279.0 mm, while October showed the lowest monthly average rainfall, with 0.0 mm) for the investigation year 2017 (Figure 1). The seedlings were planted in the ground at a distance of 2.20 m in the row, in a stony sloping terrain, managed as a meadow for more than a century. Irrigation was made by drip-wing system when occurred. The plants were not fertilized.

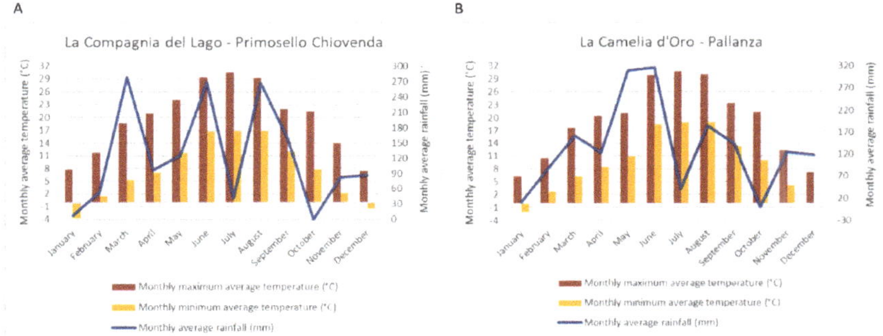

Figure 1. Monthly average temperatures (maximum and minimum— °C), and monthly average rainfall (mm) for the investigation year 2017 in: (**A**). the "Compagnia del Lago" site of cultivation. (**B**). the "Camelia d'Oro" site of cultivation.

The "Camelia d'Oro" plantation is located in Pallanza (VB) (45°56′47.9″ N 8°33′12.6″ E), where the annual average maximum temperature was 19.2 °C, the annual average minimum temperature was 9.2 °C, and the annual average rainfall was 135.6 mm for the investigation year 2017 (Figure 1). Here, before planting the seedlings, the soil was tilled with a milling cutter and a bottom fertilization with organic manure burial powder (Humus Vita, Fomet Spa, San Pietro di Morubio (VR), Italy) (25 kg per 100 m^2. N, P, and K were all present as 3–4% of the total amount, accounting on average for 8.75 g of each per m^2. Organic matter accounted for 38–45%, thus on average 103.75 g per m^2 of it were added.) was made. The plants were transferred to the ground, at a distance of 0.8 m in a row and 1.2 m in the inter-row, resulting in 1 plant per m^2. The hole in the ground was made with a drill and 2.5 L of Blond peat were added to each plant as a soil improver. The plants were tamped down and mulched with an organic mulching consisting in a mix of wood chips and leaves. In May, a cover fertilization with an organic ox-blood fertilizer (10 cc per plant) (Biostan, Aifar Agrochimica Srl, Ronco Scrivia (GE), Italy) was performed. Plants were irrigated with drip-wing irrigation with about 1.6 L of water per plant per day.

2.2. Tea Harvest and Preparation

In both plantations, fresh tea shoots with one (late spring or September harvest) or two (spring harvest between late April and early May) tender leaves and a bud were harvested.

After harvesting, tea leaves were steamed for about 1–2 min; once dried, the leaves were roasted and rolled on a hot pan at about 80 °C for about 10–15 min to stop the fermentation. They were then left in a dryer at 40 °C for 18 h, and finally stored in plastic containers at room temperature.

Thus, green teas were obtained from dried leaves received from: "La Camelia d'Oro" plantation (TCO—Tea Camellia d'Oro), providing us with samples of one shoot with one leaf (first harvest), and "La Compagnia del Lago" plantation (TCL—Tea Compagnia del Lago), providing us with a mixture of the two harvests.

2.3. *Tea Extract Preparation*

The dried leaves were ground with a mortar and pestle into a fine powder. Two hundred milliliter of deionized water was heated to 100 °C. One gram of dried tea leaves powder was added to the cooling water and left to infuse for 10 min, being stirred every two minutes. The infusion was filtered with paper filters (Whatman filter papers No. 1, Whatman, Maidstone, UK), and then with polytetrafluoroethylene (PTFE, VWR International, Milano, Italy) filters, with a 25 mm diameter and 0.45 μm pore size.

Both infusions were diluted with deionized water to obtain the working solution and maintained at −20 °C for the following analysis.

2.4. *Bioactive Compounds*

2.4.1. Total Polyphenols

The total phenolic content of diluted TCL (dilution 1:1 = 1 mL deionized water: 1 mL infusion) and TCO (dilution 1:2 = 1.2 mL deionized water: 600 μL infusion) was determined following the Folin-Ciocalteu method, as indicated by Singleton et al. [30]. The analysis was performed as follows: 1000 μL of diluted 1:10 Folin reagent were mixed with 200 μL of infusion in each plastic tube. The samples were left in the dark at room temperature for 10 min, then 800 μL of Na_2CO_3 (7.5%) were added to each tube. Samples were left in the dark at room temperature for 30 min. Absorbance was then measured at 765 nm by means of a spectrophotometer (Cary 60 UV-Vis, Agilent Technologies, Santa Clara, CA, USA), and the results were expressed in milligrams of gallic acid equivalents per 100 g of dry weight (mg GAE 100 g^{-1} DW). Analysis was performed in six replicates per infusion type.

2.4.2. Antioxidant Activity

FRAP Assay

The first procedure adopted to evaluate the antioxidant activity of diluted TCL (1:2 = 1.2 mL deionized water: 600 μL infusion) and TCO (1:3 = 1.5 mL deionized water: 500 μL infusion) was the ferric ion reducing antioxidant power (FRAP) method [31].

The FRAP solution was obtained by mixing a buffer solution at pH 3.6 ($C_2H_3NaO_2$*$3H_2O$ + $C_2H_4O_2$ in water), 2,4,6-tripyridyltriazine (TPTZ, 10 mM in HCl 40 mM), and $FeCl_3$*$6H_2O$ (20 mM).

The antioxidant activity was determined mixing 30 μL of diluted infusion with 90 μL of deionized water and 900 μL of FRAP reagent. The samples were then placed at 37 °C for 30 min. Absorbance was measured at 595 nm by means of a spectrophotometer (Cary 60 UV-Vis, Agilent Technologies, Santa Clara, CA, USA). The antioxidant activity was plotted against a $FeSO_4$*$7H_2O$ calibration curve. Extraction solution (water) without infusion was used as a control sample. Results were expressed as millimoles of ferrous iron equivalents per kilogram (mmol Fe^{2+} kg^{-1} DW). Analysis was performed in six replicates per infusion type.

DPPH Assay

The second procedure was the 2,2-diphenyl-1-picrylhydrazyl (DPPH) radical scavenging method [32].

The working solution of DPPH radical cation (DPPH˙, 100 μM) was obtained by dissolving 2 mg of DPPH in 50 mL of MeOH. The solution must have an absorbance of

1000 (±0.05) at 515 nm. To prepare the samples, 40 μL of diluted infusion were mixed with 3 mL of DPPH˙. Samples were then left in the dark at room temperature for 30 min. Absorbance was measured at 515 nm by means of a spectrophotometer (Cary 60 UV-Vis, Agilent Technologies, Santa Clara, CA, USA). The DPPH radical-scavenging activity was calculated as:

$$[(Abs0 - Abs1)/Abs0] \times 100$$

where Abs0 is the absorbance of the control (extraction solution without infusion) and Abs1 is the absorbance of the sample. The antioxidant capacity was plotted against a Trolox calibration curve and results were expressed as μmol of Trolox equivalents per gram of dry weight (μmol TE g-1 DW). Analysis was performed in six replicates per infusion type.

ABTS Assay

The third procedure was the 2,20-azino-bis-(3-ethylbenzothiazoline-6-sulphonic acid) (ABTS) method [33].

The working solution of ABTS radical cation (ABTS˙) was obtained by the reaction of 7.0 mM ABTS stock solution with 2.45 mM potassium persulfate ($K_2S_2O_8$) solution. After incubating for 12–16 h in the dark at room temperature, the working solution was diluted with distilled water to obtain an absorbance of 0.7 (±0.02) at 734 nm. The antioxidant activities of the diluted TCL (1:2) and TCO (1:3) infusions were assessed mixing 30 μL of infusion with 2 mL of diluted ABTS˙. The samples were left in the dark for 10 min. Absorbance was then measured at 734 nm by means of a spectrophotometer (Cary 60 UV-Vis, Agilent Technologies, Santa Clara, CA, USA). The ABTS radical-scavenging activity was calculated as:

$$[(Abs0 - Abs1)/Abs0] \times 100$$

where Abs0 is the absorbance of the control (extraction solution without infusion) and Abs1 is the absorbance of the sample. The antioxidant capacity was plotted against a Trolox calibration curve and results were expressed as μmol of Trolox equivalents per gram of dry weight (μmol TE g-1 DW). Analysis was performed in six replicates per infusion type.

2.4.3. Identification and Quantification of Bioactive Compounds by HPLC

The bioactive compounds contained in the infusions were determined by means of two high performance liquid chromatography-diode array detection (HPLC–DAD) method [34], using an Agilent 1200 High-Performance Liquid Chromatograph coupled to an Agilent UV-Vis diode array detector (Agilent Technologies, Santa Clara, CA, USA). Detailed chromatographic methods are reported in Table 1. Phytochemical separation was performed with a Kinetex C18 column (4.6 × 150 mm, 5 μm, Phenomenex, Torrance, CA, USA), using different mobile phases for compound identification and recording of UV spectra at different wavelengths, based on HPLC methods, as previously tested and validated [35,36] with some modifications. UV spectra were recorded at 330 nm and 280 nm. The following bioactive compounds were determined: cinnamic acids (caffeic acid, chlorogenic acid, coumaric acid, ferulic acid), flavonols (hyperoside, isoquercitrin, quercetin, quercitrin, rutin), benzoic acids (ellagic acid, gallic acid), catechins (catechin, epicatechin). All single compounds were identified by a comparison and combination of their retention times and UV spectra with those of authentic standards under the same chromatographic conditions. Results were expressed as mg 100 g^{-1} of dry weight (DW).

Table 1. HPLC methods and relative chromatographic conditions.

Method	Classes of Interest	Stationary Phase	Mobile Phase	Wavelength (nm)
A	Cinnamic acids Flavonols	KINETEX-C18 column (4.6 × 150 mm, 5 μm)	A: 10 mM KH_2PO_4/H_3PO_4 pH = 2.8 B: CH_3CN	330
B	Benzoic acids Catechins	KINETEX-C18 column (4.6 × 150 mm, 5 μm)	A: $H_2O/CH_3OH/HCOOH$ (5:95:0.1 *v/v/v*), pH = 2.5 B: $CH_3OH/HCOOH$ (100:0.1 *v/v*)	280

Elution conditions. Method A, gradient analysis: 5% B to 21% B in 17 min + 21% B in 3 min (2 min conditioning time); flow: 1.5 mL min^{-1}; Method B, gradient analysis: 3% B to 85% B in 22 min + 85% B in 1 min (2 min conditioning time); flow: 0.6 mL min^{-1}.

2.5. Statistical Analysis

All data were subjected to the statistical analysis for the normality and homoscedasticity through Saphiro-Wilk's test ($p > 0.05$) and Levene's test ($p < 0.05$), respectively. Mean comparisons were computed using the SPSS 25 software (version 25.0; SPSS Inc., Chicago, IL, USA). Correlations among the bioactive compounds of the two teas were calculated by Pearson's correlation coefficient test by means of PAST 4.03 software. Principal coordinate analysis (PCA)—Biplot was performed using the same software. Eigenvalues were calculated using a covariance matrix among 30 traits as input, and the two-dimensional PCA biplot was constructed.

3. Results

The total polyphenol content and the antioxidant activity (FRAP, DPPH, and ABTS assays) of both examined teas are reported in Table 2.

Table 2. Total polyphenols, and antioxidant activity of Tea Compagnia del Lago (TCL) and Tea Camellia d'Oro (TCO). Data are presented as mean value ± standard deviation.

Tea Type	Total Polyphenols (mg GAE/100 g DW)	Antioxidant Activity				
		FRAP (mmol Fe^{2+}/kg)	DPPH		ABTS	
			(μmol TE/g DW)	Inhibition %	(μmol TE/g DW)	Inhibition %
TCL	8540.42 ± 105.38	25796.61 ± 951.83	302.35 ± 10.4	46.9	623.44 ± 4.64	94.3
TCO	14918.91 ± 222.31	42758.86 ± 933.85	532.37 ± 5.95	61.2	881.08 ± 1.81	99.8
p	**	***	***		***	

The statistical relevance is provided (** = $p < 0.01$; *** = $p < 0.001$).

The total phenolic content varied from 8540.42 to 14918.91 mg GAE 100 g^{-1} being significantly higher in TCO. The antioxidant activity resulted higher for TCO rather than for TCL (42758.86 and 25796.61 mmol Fe^{2+} kg^{-1} for the FRAP assay, respectively; 532.37 and 302.35 μmol TE g^{-1} DW for the DPPH assay, respectively; 881.08 and 623.44 μmol TE g^{-1} DW for the ABTS assay, respectively).

In the TCO infusion, 13 compounds out of 13 were found by means of HPLC analysis, namely caffeic acid, chlorogenic acid, coumaric acid, ferulic acid, hyperoside, isoquercitrin, quercetin, quercitrin, rutin, ellagic acid, gallic acid, catechin, and epicatechin, while in TCL infusion, only 9 compounds out of 13 were found (chlorogenic acid, coumaric acid, ferulic acid, and quercetin were not detected) (Figure 2, Table 3).

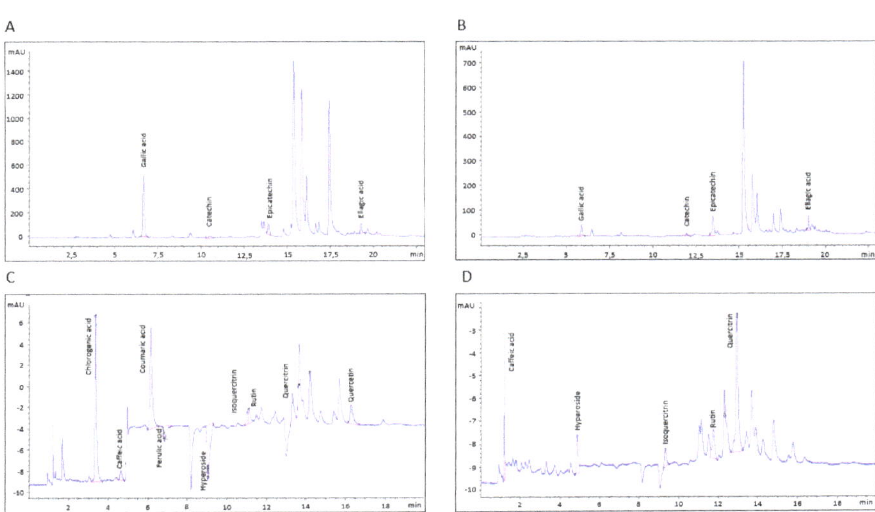

Figure 2. Chromatographic profile of benzoic acids and catechins (**A**), and cinnamic acids and flavonols (**C**) in TCO. Chromatographic profile of benzoic acids and catechins (**B**) and cinnamic acids and flavonols (**D**) in TCL. TCO = Tea Camellia d'Oro. TCL = Tea Compagnia del Lago.

Table 3. Bioactive compounds in Tea Compagnia del Lago (TCL) and Tea Camellia d'Oro (TCO). Data are presented as mean value ± standard deviation. The values are given in mg 100 g^{-1} DW.

Tea type	Flavonols				
	Hyperoside	Isoquercitrin	Quercetin	Quercitrin	Rutin
TCL	25.37 ± 4.39	35.46 ± 2.60	-	242.38 ± 10.11	44.66 ± 3.93
TCO	28.24 ± 4.30	35.31 ± 4.38	388.28 ± 95.47	113.11 ± 14.75	42.32 ± 4.19
p	ns	ns		***	ns
Tea type	**Cinnamic acids**				
	Caffeic acid	**Chlorogenic acid**	**Coumaric acid**	**Ferulic acid**	
TCL	42.57 ± 6.67	-	-	-	
TCO	43.36 ± 1.72	612.25 ± 37.58	204.62 ± 16.47	57.85 ± 13.32	
p	ns				
Tea type	**Benzoic acids**		**Catechins**		
	Ellagic acid	**Gallic acid**	**Catechin**	**Epicatechin**	
TCL	59.06 ± 2.33	42.39 ± 2.37	122.06 ± 10.86	770.39 ± 21.06	
TCO	86.85 ± 5.17	803.88 ± 56.99	478.98 ± 27.53	735.84 ± 89.76	
p	**	***	***	ns	

The statistical relevance is provided (ns = non-significant; ** = $p < 0.01$; *** = $p < 0.001$). -: compound not detected.

The two teas showed significant different contents in quercitrin (242.38 mg 100 g^{-1} DW in TCL and 113.11 mg 100 g^{-1} DW in TCO). TCO and TCL showed significant different contents in cinnamic acids, with TCO being richer in these compounds than TCL. Benzoic acids and catechins were significantly higher in TCO than in TCL (except for epicatechin, which showed no significant differences in the two infusions).

The phenolic content in both teas proved to be significantly correlated to their antioxidant activity (Figure 3), for the three methods used (FRAP, DPPH, ABTS). The phenolic content was also positively correlated ($p < 0.05$) to chlorogenic acid, coumaric acid, ferulic acid, quercetin, ellagic acid, gallic acid, and catechin content. Interestingly, chlorogenic acid was found to be negatively correlated ($p < 0.05$) to the quercitrin content, and this latter was negatively correlated to gallic acid too.

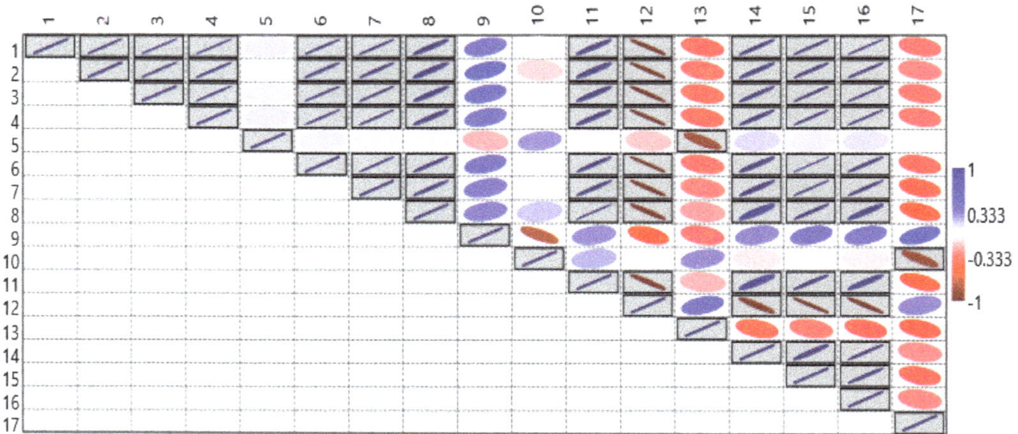

Figure 3. Pearson correlation among the bioactive compounds of Tea "Compagnia del Lago" (TCL) and Tea "Camelia d'Oro" (TCO). Boxed ellipses are significantly correlated ($p < 0.05$). Blue indicates positive correlation and red indicates negative correlation. The more the ellipse is compressed, the stronger the correlation. 1. Polyphenols; 2. FRAP; 3. DPPH; 4. ABTS; 5. caffeic acid; 6. chlorogenic acid; 7. coumaric acid; 8. ferulic acid; 9. hyperoside; 10. isoquercitrin; 11. quercetin; 12. quercitrin; 13. rutin; 14. ellagic acid; 15. gallic acid; 16. catechin; 17. epicatechin.

The relationship between the studied parameters were evaluated through a PCA and represented in a two-dimensional PCA scatterplot (based on the first two principal components (PCs)) (Figure 4). As depicted, the first two PCs explained 77.8% of total variation. The first PC accounted for 62.2% and was positively correlated with polyphenols, antioxidant activity (FRAP, DPPH, and ABTS), caffeic acid, chlorogenic acid, coumaric acid, ferulic acid, hyperoside, isoquercitrin, quercetin, ellagic acid, gallic acid, and catechin; conversely, it was negatively correlated with quercitrin, rutin, and epicatechin. The second PC accounted for 15.6% and was positively correlated with chlorogenic acid, coumaric acid, ferulic acid, isoquercitrin, quercetin, quercitrin, rutin, gallic acid; conversely, it was negatively correlated with polyphenols, antioxidant activity (FRAP, DPPH, and ABTS), caffeic acid, hyperoside, ellagic acid, catechin, and epicatechin. The scatterplot showed that the two teas are clearly distinguished, confirming that TCO is mainly linked to almost all the detected bioactive compounds. In the opposite, TCL was correlated only to quercitrin, rutin, and epicatechin.

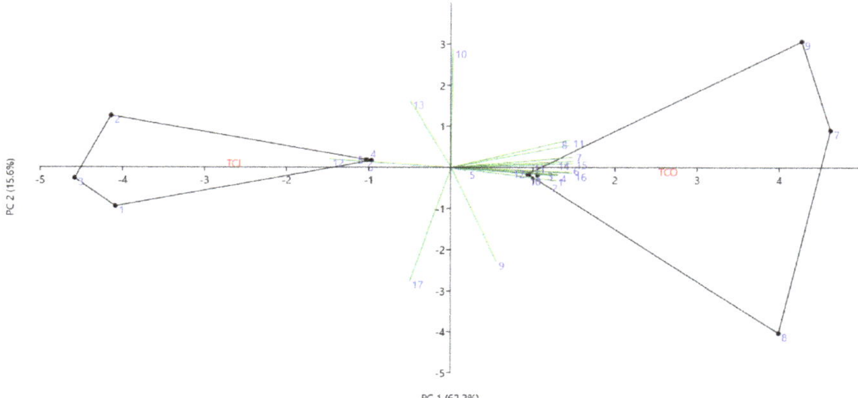

Figure 4. Principal component analysis (PCA)-biplot of Tea Compagnia del Lago (TCL) and Tea Camelia d'Oro (TCO), by means of the PAST 4.03 software. 1. Polyphenols; 2. FRAP; 3. DPPH; 4. ABTS; 5. caffeic acid; 6. chlorogenic acid; 7. coumaric acid; 8. ferulic acid; 9. hyperoside; 10. isoquercitrin; 11. quercetin; 12. quercitrin; 13. rutin; 14. ellagic acid; 15. gallic acid; 16. catechin; 17. epicatechin.

4. Discussion

The Italian tea "Camelia d'Oro" (TCO) showed a total phenolic content (14918.91 mg GAE 100 g^{-1} DW) higher than the other green teas analyzed in similar studies. Cai et al. [37] purchased the green tea sample in a Chinese trade market, finding a total phenolic content of 13,600 mg GAE 100 g^{-1} DW. Mildner-Szkudlarz et al. [38] obtained Yunan green tea leaves in a special tea store, then ground leaves were extracted for 24 h using a 95% ethanol solution; the analysis showed a total phenolic content of 13,654 mg GAE 100 g^{-1} DW. TCO even showed a higher phenolic content than the white tea analyzed by Zielinski et al. [39], which obtained a wide range: 3800–11,400 mg GAE 100 g^{-1} DW. Conversely, the "Compagnia del Lago" tea (TCL) showed a total phenolic content of 8540.42 ± 105.38 mg GAE 100 g^{-1} DW, lower than the previous studies' results. These data therefore might indicate that the cultivation and processing methods adopted by "La Camelia d'Oro" plantation promoted the accumulation of phenolic substances. However, Pérez-Burillo et al. [8] found a higher polyphenols content in commercial green tea in Spain, brewed in water for 7 min at 98 °C, obtaining a value of 1043 ± 5 mg GAE L^{-1} (TCO = 745.95 mg GAE L^{-1}).

The antioxidant activity of the two teas, assessed with the DPPH, ABTS, and FRAP methods, confirmed the phenolic trend, being higher in TCO than in TCL. Regarding to the DPPH assay, TCO and TCL showed values of 532.37 µmol TE g^{-1} DW and 302.35 µmol TE g^{-1} DW respectively, with an average inhibition percentage of 46.9% and 61.2% respectively. These results are higher than those published by Sirichaiwetchakoon et al. [3], who found an inhibition percentage of 41.46% for commercial green tea, purchased in a supermarket in the United Kingdom and prepared by boiling ground tea leaves in 80 °C 1x phosphate buffered saline (PBS) for 5 min. Conversely, Lv et al. [40] found a higher inhibition percentage in three black teas, ranging from 82.3% to 87.6%. This is unusual, as many researchers suggest that green tea has more antioxidant activity than black tea [38], as green tea has no fermentation processing [41]. However, the same authors [40] found a lower inhibition percentage through the ABTS assay, ranging from 12.08 to 18.08%, according to the black tea type. Conversely, in this study the ABTS assay confirmed the DPPH high values for both tea samples, ranging from 94.3% in TCL to 99.8% in TCO, thus confirming the highest antioxidant activity of green tea.

A few works also evaluated the antioxidant activity by means of FRAP test, however, using different units of measurement, making it complex to compare the different values [8,9].

The "Camelia d'Oro" tea (TCO) resulted to contain all the 13 bioactive compounds investigated, while the "Compagnia del Lago" tea (TCL) showed only 9. The major components identified in both Italian teas were quercetin, catechin, and epicatechin; in TCO also chlorogenic acid and gallic acid. Although not giving the single related values, Cai et al. [37] also found that catechin, epicatechin, and quercetin were the major types of phenolic compounds found in tea leaves.

The "Camelia d'Oro" tea (TCO) showed catechins (478.98 mg 100 g^{-1} DW) and epicatechins (735.84 mg 100 g^{-1} DW) values (Table 3) similar or higher than other green teas. Li et al. [9] found in the green Longjing tea values of 0.36 mg 100 mg^{-1} DW for catechins and 0.90 mg 100 mg^{-1} DW for epicatechins. Hyun et al. [25] found lower catechin values (ranging from 3.14 to 4.76 mg g^{-1} FW), but higher epicatechin values (ranging from 8.12 to 10.40 mg g^{-1} FW) than TCO. Conversely, the "Compagnia del Lago" tea (TCL) showed catechins (122.06 mg 100 g^{-1} DW) and epicatechins (770.39 mg 100 g^{-1} DW) values lower than the previous studies results.

It is noteworthy that the catechin content is highly correlated to the antioxidant activity measured with the three assays (FRAP, DPPH, and ABTS) (Table 3), as already found by Pérez-Burillo et al. [8], when comparing white and green teas. Catechins are known to have important human health benefits, due to their antioxidative, anti-inflammatory, anticarcinogenic, antidiabetic, and antimicrobial properties [34,41], and they may also help reduce the body mass index [42]. Quercetin shows antihypertensive effects, improving endothelial function, gene expression, and modulating cell signals [43], while coumaric acid has protective effects against carcinogenesis, atherosclerosis, oxidative cardiac damages and has anti-inflammatory effects [44].

Gallic acid of TCO (803.88 mg 100 g^{-1} DW) was higher than the Longjing tea values (0.07 mg 100 mg^{-1} DW) [9], conversely TCL (42.39 mg 100 g^{-1} DW) values were lower.

Although not showing the single related values, Gorjanović et al. [45] showed the presence of ferulic and caffeic acids in green tea infusions, as they were detected in this study (Table 3).

Flavonols (hyperoside, isoquercitrin, quercitrin, rutin), chlorogenic acid, and ellagic acid were not frequently reported individually in other studies, but more as a single category of flavonols, cinnamic acids, and benzoic acids; they were almost all found in both the Italian teas.

Flavonols are health-promoted compounds [42] which can be largely found in green tea leaves [46], and chlorogenic acid has antioxidant, anti-inflammatory, anticancer, antilipidemic, antidiabetic, antihypertensive, and antineurodegenerative activities [47], thus these teas are a rich source of bioactive compounds.

As already observed by Li et al., Lv et al., and Tenore et al. [9,21,40], antioxidant activity was positively correlated with polyphenol content and, thus, with catechins content. The methods for the antioxidant activity analysis were also positively correlated with each other, as Zielinski et al. [39] observed.

5. Conclusions

This work investigated the content in bioactive compounds of two Italian teas, both cultivated in the Lake Maggiore District, in Piedmont. The results indicated that the two teas have different phenolic compositions, probably due to the different cultivation substrates or techniques.

The "Camelia d'Oro" tea (TCO) showed in general, higher bioactive compounds levels than the "Compagnia del Lago" tea (TCL).

However, both teas showed values in accordance with other studies' results, or even higher, confirming that they would be suitable to diversify the Italian growers' production with the *C. sinensis* cultivation.

Author Contributions: Conceptualization, V.S.; methodology, M.C.; investigation, N.M.F., S.D., M.C.; data curation, N.M.F., M.C.; writing—original draft preparation, N.M.F.; writing—review and editing, M.C., S.D., V.S.; supervision, V.S.; project administration, V.S.; funding acquisition, V.S. All authors have read and agreed to the published version of the manuscript.

Funding: This research received no external funding.

Institutional Review Board Statement: Not applicable.

Informed Consent Statement: Not applicable.

Data Availability Statement: Not applicable.

Acknowledgments: Authors acknowledge Dario Donno for HPLC analysis, and the two nurseries, "La Camellia d'Oro" and "La Compagnia del Lago" for providing the two tea leaves.

Conflicts of Interest: The authors declare no conflict of interest.

References

1. Komes, D.; Horžić, D.; Belščak, A.; Ganić, K.K.; Vulić, I. Green tea preparation and its influence on the content of bioactive compounds. *Food Res. Int.* **2010**, *43*, 167–176. [CrossRef]
2. Ori, F.; Ma, J.; Gori, M.; Lenzi, A.; Chen, L.; Giordani, E. DNA-based diversity of tea plants grown in Italy. *Genet. Resour. Crop Evol.* **2017**, *64*, 1905–1915. [CrossRef]
3. Sirichaiwetchakoon, K.; Lowe, G.M.; Eumkeb, G. The Free Radical Scavenging and Anti-Isolated Human LDL Oxidation Activities of *Pluchea indica* (L.) Less. Tea Compared to Green Tea (*Camellia sinensis*). *Biomed. Res. Int.* **2020**, *2020*, 1–12. [CrossRef]
4. Koo, S.I.; Noh, S.K. Green tea as inhibitor of the intestinal absorption of lipids: Potential mechanism for its lipid-lowering effect. *J. Nutr. Biochem.* **2007**, *18*, 179–183. [CrossRef] [PubMed]
5. Xing, L.; Zhang, H.; Qi, R.; Tsao, R.; Mine, Y. Recent Advances in the Understanding of the Health Benefits and Molecular Mechanisms Associated with Green Tea Polyphenols. *J. Agric. Food Chem.* **2019**, *67*, 1029–1043. [CrossRef] [PubMed]
6. Sharangi, A.B. Medicinal and therapeutic potentialities of tea (*Camellia sinensis* L.)—A review. *Food Res. Int.* **2009**, *42*, 529–535. [CrossRef]
7. Jiang, X.; Liu, Y.; Li, W.; Zhao, L.; Meng, F.; Wang, Y.; Tan, H.; Yang, H.; Wei, C.; Wan, X.; et al. Tissue-Specific, Development-Dependent Phenolic Compounds Accumulation Profile and Gene Expression Pattern in Tea Plant [*Camellia sinensis*]. *PLoS ONE* **2013**, *8*, e62315. [CrossRef] [PubMed]
8. Pérez-burillo, S.; Giménez, R.; Ru, J.A.; Pastoriza, S. Effect of brewing time and temperature on antioxidant capacity and phenols of white tea: Relationship with sensory properties. *Food Chem.* **2018**, *248*, 111–118. [CrossRef]
9. Li, K.; Shi, X.; Yang, X.; Wang, Y.; Ye, C.; Yang, Z. Antioxidative activities and the chemical constituents of two Chinese teas, *Camellia kucha* and *C. ptilophylla*. *Int. J. Food Sci. Technol.* **2012**, *47*, 1063–1071. [CrossRef]
10. Food and Agriculture Organization of the United Nations. Current Market Situation and Medium Term Outlook. In Proceedings of the Twenty-Third Session of the Intergovernmental Group on Tea, Hangzhou, China, 17–20 May 2018; pp. 13–16.
11. FAO. Emerging Trends in Tea Consumption: Informing a Generic Promotion Process. In Proceedings of the Twenty-Third Session of the Intergovernmental Group on Tea, Hangzhou, China, 17–20 May 2018; pp. 1–9.
12. Food and Agriculture Organization of the United Nations. Developing futures and swap markets for tea. In Proceedings of the Twenty-Third Session of the Intergovernmental Group on Tea, Hangzhou, China, 17–20 May 2018; pp. 1–6.
13. Bhardwaj, P.; Kumar, R.; Sharma, H.; Tewari, R.; Ahuja, P.S.; Sharma, R.K. Development and utilization of genomic and genic microsatellite markers in Assam tea (*Camellia assamica* ssp. *assamica*) and related Camellia species. *Plant Breed.* **2013**, *132*, 748–763. [CrossRef]
14. Khan, N.; Mukhtar, H. Tea polyphenols for health promotion. *Life Sci.* **2007**, *81*, 519–533. [CrossRef] [PubMed]
15. Zheng, Q.; Li, W.; Zhang, H.; Gao, X.; Tan, S. Optimizing synchronous extraction and antioxidant activity evaluation of polyphenols and polysaccharides from Ya'an Tibetan tea (*Camellia sinensis*). *Food Sci. Nutr.* **2019**, *8*, 489–499. [CrossRef] [PubMed]
16. Heber, D.; Zhang, Y.; Yang, J.; Ma, J.E.; Henning, S.M.; Li, Z. Green tea, black tea, and oolong tea polyphenols reduce visceral fat and inflammation in mice fed high-fat, high-sucrose obesogenic diets. *J. Nutr.* **2014**, *144*, 1385–1393. [CrossRef]
17. Senanayake, S.P.J.N. Green tea extract: Chemistry, antioxidant properties and food applications—A review. *J. Funct. Foods* **2013**, *5*, 1529–1541. [CrossRef]
18. Karori, S.M.; Wachira, F.N.; Wanyoko, J.K.; Ngure, R.M. Antioxidant capacity of different types of tea products. *African J. Biotechnol.* **2007**, *6*, 2287–2296. [CrossRef]
19. Afzal, M.; Safer, A.M.; Menon, M. Green tea polyphenols and their potential role in health and disease. *Inflammopharmacology* **2015**, *23*, 151–161. [CrossRef]
20. Hayakawa, S.; Ohishi, T.; Miyoshi, N.; Oishi, Y.; Nakamura, Y.; Isemura, M. Anti-Cancer Effects of Green Tea Epigallocatchin-3-Gallate and Coffee Chlorogenic Acid. *Molecules* **2020**, *25*, 4553. [CrossRef]

21. Tenore, G.C.; Daglia, M.; Ciampaglia, R.; Novellino, E. Exploring the Nutraceutical Potential of Polyphenols from Black, Green and Exploring the Nutraceutical Potential of Polyphenols from Black, Green and White Tea Infusions—An Overview. *Curr. Pharm. Biotechnol.* **2015**, *16*, 265–271. [CrossRef] [PubMed]
22. Huo, C.; Dou, Q.P.; Chan, T.H. Synthesis of phosphates and phosphates-acetates hybrids of green tea polyphenol (-)-epigallocatechine-3-gallate (EGCG) and its G ring deoxy analogs as potential anticancer prodrugs. *Tetrahedron Lett.* **2011**, *52*, 5478–5483. [CrossRef]
23. Kellogg, J.J.; Graf, T.N.; Paine, M.F.; McCune, J.S.; Kvalheim, O.M.; Oberlies, N.H.; Cech, N.B. Comparison of Metabolomics Approaches for Evaluating the Variability of Complex Botanical Preparations: Green Tea (*Camellia sinensis*) as a Case Study. *J. Nat. Prod.* **2017**, *80*, 1457–1466. [CrossRef]
24. Bramel, P.J.; Chen, L. *A Global Strategy for the Conservation and Use of Tea Genetic Resources*; Crop Trust: Bonn, Germany, 2019.
25. Hyun, D.Y.; Gi, G.Y.; Sebastin, R.; Cho, G.T.; Kim, S.H.; Yoo, E.; Lee, S.; Son, D.M.; Lee, K.J. Utilization of phytochemical and molecular diversity to develop a target-oriented core collection in tea germplasm. *Agronomy* **2020**, *10*, 1667. [CrossRef]
26. Camangi, F.; Stefani, A.; Bracci, T.; Minnocci, A.; Sebastiani, L.; Lippi, A.; Cattolica, G.; Santoro, A.M. *Antiche Camelie Della Lucchesia*; Sant'Anna, S.S., Ed.; ETS: Pisa, Italy, 2012; ISBN 978884673128-9.
27. Caser, M.; Berruti, A.; National, I.; Bianciotto, V.; National, I.; Devecchi, M. Floriculture and territory—The protection of the traditional Italian tipicity: The case of "La Camelia del Lago Maggiore (PGI)". *Acta Hortic.* **2018**. [CrossRef]
28. Corneo, A.; Remotti, D. *Camelie Dell'ottocento Nel Verbano*; Regione Piemonte: Torino, Italy, 2000.
29. Freda, R.; Borrello, M.; Cembalo, L. Innovation in Floriculture When Environmental and Economics criteria are conflicting. *Calitatea* **2015**, *16*, 110–118.
30. Singleton, V.L.; Orthofer, R.; Lamuela-Raventós, R.M. Analysis of total phenols and other oxidation substrates and antioxidants by means of Folin- Ciocalteu reagent. *Methods Enzymol.* **1999**, *299*, 152–178.
31. Benzie, I.F.F.; Strain, J.J. The ferric reducing ability of plasma (FRAP) as a measure of "antioxidant power": The FRAP assay. *Anal. Biochem.* **1996**, *239*, 70–76. [CrossRef]
32. Wong, S.P.; Leong, L.P.; William Koh, J.H. Antioxidant activities of aqueous extracts of selected plants. *Food Chem.* **2006**, *99*, 775–783. [CrossRef]
33. Urbani, E.; Blasi, F.; Stella, M.; Claudia, S.; Cossignani, L. Investigation on secondary metabolite content and antioxidant activity of commercial saffron powder. *Eur. Food Res. Technol.* **2016**, *242*, 987–993. [CrossRef]
34. Caser, M.; Demasi, S.; Stelluti, S.; Donno, D.; Scariot, V.; Crocus sativus, L. Cultivation in Alpine Environments: Stigmas and Tepals as Source of Bioactive Compounds. *Agronomy* **2020**, *10*, 1473. [CrossRef]
35. Caser, M.; Demasi, S.; Victorino, M.M.I.; Donno, D.; Faccio, A.; Lumini, E.; Bianciotto, V.; Scariot, V. Arbuscular Mycorrhizal Fungi Modulate the Crop Performance and Metabolic Profile of Saffron in Soilless Cultivation. *Agronomy* **2019**, *9*, 232. [CrossRef]
36. Donno, D.; Mellano, M.G.; Riondato, I.; De Biaggi, M.; Andriamaniraka, H.; Gamba, G.; Beccaro, G.L. Traditional and Unconventional Dried Fruit Snacks as a Source of Health-Promoting Compounds. *Antioxidants* **2019**, *8*, 396. [CrossRef] [PubMed]
37. Cai, Y.; Luo, Q.; Sun, M.; Corke, H. Antioxidant activity and phenolic compounds of 112 traditional Chinese medicinal plants associated with anticancer. *Life Sci.* **2004**, *74*, 2157–2184. [CrossRef] [PubMed]
38. Mildner-Szkudlarz, S.; Zawirska-Wojtasiak, R.; Obuchowskij, W.; Goslinski, M. Evaluation of Antioxidant Activity of Green Tea Extract and Its Effect on the Biscuits Lipid Fraction Oxidative Stability. *Sens. Food Qual.* **2009**, *74*, 362–370. [CrossRef]
39. Zielinski, A.A.F.; Haminiuk, C.W.I.; Beta, T. Multi-response optimization of phenolic antioxidants from white tea (*Camellia sinensis* L. Kuntze) and their identification by LC-DAD-Q-TOF-MS/MS. *LWT Food Sci. Technol.* **2016**, *65*, 897–907. [CrossRef]
40. Lv, H.; Zhang, Y.; Shi, J.; Lin, Z. Phytochemical profiles and antioxidant activities of Chinese dark teas obtained by different processing technologies. *Food Res. Int.* **2017**, *100*, 486–493. [CrossRef]
41. Ananingsih, V.K.; Sharma, A.; Zhou, W. Green tea catechins during food processing and storage: A review on stability and detection. *FRIN* **2013**, *50*, 469–479. [CrossRef]
42. Demasi, S.; Caser, M.; Donno, D.; Enri, S.R.; Lonati, M.; Scariot, V. Exploring wild edible flowers as a source of bioactive compounds: New perspectives in horticulture. *Folia Hortic.* **2021**, *33*, 1–22. [CrossRef]
43. Xue, Z.; Wang, J.; Chen, Z.; Ma, Q.; Guo, Q.; Gao, X.; Chen, H. Antioxidant, antihypertensive, and anticancer activities of the flavonoid fractions from green, oolong, and black tea infusion waste. *J. Food Biochem.* **2018**, *42*. [CrossRef]
44. Cui, P.; Zhong, W.; Qin, Y.; Tao, F.; Wang, W.; Zhan, J. Characterization of two new aromatic amino acid lyases from actinomycetes for highly efficient production of p—Coumaric acid. *Bioprocess Biosyst. Eng.* **2020**, *43*, 1287–1298. [CrossRef]
45. Gorjanović, S.; Komes, D.; Pastor, F.T.; Belşçak-Cvitanović, A.; Pezo, L.; Heçimović, I.; Suznjević, D. Antioxidant capacity of teas and herbal infusions: Polarographic assessment. *J. Agric. Food Chem.* **2012**, *60*, 9573–9580. [CrossRef] [PubMed]
46. Lee, M.K.; Kim, H.-W.; Lee, S.-H.; Kim, Y.J.; Asamenew, G.; Choi, J.; Lee, J.-W.; Jung, H.-A.; Yoo, S.M.; Kim, J.-B. Characterization of catechins, theaflavins, and flavonols by leaf processing step in green and black teas (*Camellia sinensis*) using UPLC-DAD-QToF/MS. *Eur. Food Res. Technol.* **2019**, *245*, 997–1010. [CrossRef]
47. Santana-Gálvez, J.; Cisneros-Zevallos, L.; Jacobo-Velázquez, D.A. Chlorogenic Acid: Recent advances on its dual role as a food additive and a nutraceutical against metabolic syndrome. *Molecules* **2017**, *22*, 358. [CrossRef] [PubMed]

 horticulturae

MDPI

Article

The Effects of Post-Harvest Treatments on the Quality of *Agastache aurantiaca* Edible Flowers

Ilaria Marchioni [1], Rosanna Dimita [2], Giovanni Gioè [1], Luisa Pistelli [3,4], Barbara Ruffoni [5], Laura Pistelli [1,4,*] and Basma Najar [3]

1 Department of Agriculture Food and Environment, University of Pisa, Via del Borghetto 80, 56124 Pisa, Italy; ilaria.marchioni.16@gmail.com (I.M.); giovanni.gioe2@gmail.com (G.G.)
2 Chambre d'Agriculture des Alpes-Maritimes (CREAM), MIN Fleurs 17 Box 85, CEDEX 3, 06296 Nice, France; dimitarosanna@gmail.com
3 Pharmacy Department, University of Pisa, Via Bonanno 6, 56126 Pisa, Italy; luisa.pistelli@unipi.it (L.P.); basmanajar@hotmail.fr (B.N.)
4 Interdepartmental Research Center "Nutraceuticals and Food for Health", University of Pisa, Via del Borghetto 80, 56124 Pisa, Italy
5 CREA—Centro di Ricerca Orticoltura e Florovivaismo, Corso Inglesi 508, 18038 Sanremo, Italy; barbara.ruffoni@crea.gov.it
* Correspondence: laura.pistelli@unipi.it; Tel.: +39-0502216536

Abstract: *Agastache* spp. are used as ornamental plants for their pleasant aroma and the different colors of flowers. Nowadays, their edible flowers have become attractive for their nutraceutical properties. Post-harvest treatment appears as a crucial point to avoid impairment of the nutraceutical compounds and aroma, so different treatments were tested to analyze their effect on the bioactive metabolites and volatilome. Results indicated that freeze-drying was the best solution to prolong the shelf life of these flowers. The use of high temperatures (50, 60, 70 °C) led to altered the composition of antioxidant compounds (phenolic compounds, flavonoids, anthocyanins, carotenoids). Air-drying at 30 °C was a reasonable method, even though time consuming. Concerning the aroma profile, all samples were dominated by oxygenated monoterpene compounds. Pulegone was the main or one of the major constituents of all samples together with *p*-menthone. Gas Chromatography-Mass Spectrometry results showed a correlation between the temperature and the number of identified compounds. Both fresh and freeze-dried samples evidenced a lesser number (10 and 19, respectively); when the temperature raised, the number of identified constituents increased. Statistical analyses highlighted significant differences between almost all aromatic compounds, even if both Principal Component and Hierarchical Cluster analyses differed at 60 and 70 °C and from the other treatments.

Keywords: carotenoids; phenolic compounds; antioxidant activity; VOCs; aroma; air-drying; freeze-drying; preservation

Citation: Marchioni, I.; Dimita, R.; Gioè, G.; Pistelli, L.; Ruffoni, B.; Pistelli, L.; Najar, B. The Effects of Post-Harvest Treatments on the Quality of *Agastache aurantiaca* Edible Flowers. *Horticulturae* **2021**, *7*, 83. https://doi.org/10.3390/horticulturae7040083

Academic Editors: Lucia Guidi, Luigi De Bellis and Alberto Pardoss

Received: 14 March 2021
Accepted: 31 March 2021
Published: 15 April 2021

1. Introduction

Edible flowers (EF) have been known for a long time [1]. EF are commonly consumed as vegetables, such as broccoli, cauliflowers, and artichokes, and are normally ascribed as horticultural products. Nowadays, the recovery of ancient and folk recipes with uncommon and refined ingredients are becoming of great interest. These new food ingredients can be included in the diet if their nutraceutical properties are properly defined. Regarding this topic, several flower varieties have entered the horticultural market, and new EF have been selected for their pleasant aroma and phytochemical composition [2–6]. The edible part of flowers consists mainly in petals and the reproductive organs, such as carpels and stamens [7]. In some cases, pollen of specific flowers may contain potential allergens [8], and susceptible people should carefully introduce new varieties of flowers into their diet, in order to check for any allergic reactions [8]. Bioactive compounds are detected in different amounts, and differ in relation to the flower anatomy, botanical family, and attitude to environmental condition [9,10]. Carbohydrates are the main constituents in EF, followed by other components with biological activities such as vitamins, minerals, and antioxidant compounds [11–15]. Phenolic compounds are the main antioxidant compounds, usually detected in the petals [11]. Phenolic acids occur naturally in plants and their intake in a diet contributes to preventing cancer along with cardiovascular and neurodegenerative diseases [11]. Similar biological activities are ascribed to flavonoids and anthocyanins, as subgroups of phenolic compounds [11]. Pigments, such as anthocyanins and carotenoids, have insect attractive functions, characterized also by important antioxidant activities [16–18].

EF are usually consumed fresh, but they are highly perishable within few days of harvest, and the postharvest period is very crucial for their longevity [19,20]. The senescence process is often mediated by ethylene and by programmed cell death. The first visible signals of deterioration are alteration of flowers morphology, color, and sometimes their aroma [21]. Many different technologies are used to prolong the cell vitality or to ensure a good maintenance of some bioactive compounds. Cold treatment is frequently used to prolong the fresh flowers' longevity up to one week, even if loss of pigmentation, vitamins content (ascorbic acid), and other metabolites such as phenolics can occur [20]. Drying is another common method used to prolong EF shelf life. This process can be achieved by using different instruments and temperatures, even though the results can be different in terms of visual quality and metabolite composition [22]. The oven/air drying method is traditionally used with herbs and spice, or flowers for infusion, but the prolonged treatment may negatively affect the nutritional properties [22,23]. Freeze-drying is the alternative strategy able to better preserve flowers' nutraceutical properties, especially the thermolabile compounds [20,24].

The Lamiaceae family consists of hundreds of genus and thousands of species, and the *Agastache* genus is one of the most representative taxa, with twenty-two herbaceous perennial aromatic species [25]. These plants are commonly used as herbal drugs and spice, as well as a source of essential oils (used in perfumes). Several *Agastache* species are cultivated as bee forage, ornamental plants, and for therapeutic purposes [25,26]. Tubular two-lipped flowers are characterized by different colors and fragrances [27,28]. The phytochemical compounds and aroma profile have been recently described for a few species [2,25].

Agastache aurantiaca (A. Gray) Lint & Epling is native to North America, but it also grows spontaneously in Mexico (Sierra Madre of southwestern Chihuahua and western Durango), distributed in rocky fields, plateaus, and canyon summits in open pine-oak woodlands [29]. To the best of our knowledge, very few studies were performed on this species, mainly focused on plant essential oils [30], flavonoids [31], anthocyanins [32], and trace elements [4]. Flowers of the variety 'Sunset Yellow' were recently investigated for the assessment of nutritional compounds, the aroma, and essential oil profiles, suggesting their use as EFs [2].

In this work, *Agastache aurantiaca* 'Apricot Sprit' was examined. The flowers have a light orange color, and their full blooming occurs from June to September [33]. The aim of this work was to investigate for the first time the nutraceutical properties (secondary metabolites with antioxidant activity and total sugars quantification) and the aroma profile of fresh and dried flowers (freeze-dried, hot-air dried at 30, 50, 60, 70 °C), in order to identify the most suitable drying temperature able to guarantee a high-quality product.

2. Materials and Methods

2.1. Plant Materials

Young plants obtained from cuttings of *Agastache aurantiaca* (A.Gray) Lint & Epling 'Apricot Sprite', were purchased online from a French nursery company, and grown in 10-L pots, filled with peat characterized by pH 6.1, electrical conductivity 0.38 dS/m, bulk density 120 kg/m^3, total porosity 94% v/v. Fertigation with a drip irrigation system was performed once a day (0.33 L per plant) with NPK mineral fertilizer and a mixture of microelements (Plantiol, New Polyplants srl, Lucca, Italy). The plants were grown until full-blooming in an unheated greenhouse at La Chambre d'Agriculture des Alpes-Maritimes (CREAM, Nice, France, GPS: 43.668318 N, 7.204194 E). High quality flowers, carefully selected without any signs of disease or aesthetic/sensory defect, were harvested and used for the postharvest treatments (reported in Post-harvest treatments). The collection of the inflorescences took place in the summer flowering period (2019), between 8:00 and 10:00 in the morning. Subsequently, the fresh flowers were weighed (FW), packaged, and frozen at a temperature of −80 °C.

2.2. Post-Harvest Treatments

Flowers of *A. aurantica* 'Apricot Sprite' were initially weighted (FW), and then vacuum freeze-dried (Labconco, Kansas City, MO, USA) for 48 h, or hot-air dried with an electric dryer (Dejelin Nutri Dry with constant ventilation). The hot-air drying process proceeded until the dry weight (DW) was unchanged (the moisture reached 16–18%, data not shown). Different temperatures were used (30, 50, 60, 70 °C) for 65, 12, 5, 3 h, respectively. The percentage of weight loss was calculated by the following formula: (FW-DW) × 100/FW. Fresh, dried, and freeze-dried flowers were analyzed at the Plant Physiology Laboratory of the Department of Agricultural, Food and Environment, University of Pisa.

2.3. Biochemical Analyses

For total carotenoids quantification, 100 mg of fresh flowers and 40 mg of dried flowers were extracted in 10 mL of pure methanol for 24 h at 4 °C. The absorbance was determined at 470 nm (UV-VIS Spectrophotometer, SHIMADZU UV-1800, Shimadzu Corp., Kyoto, Japan), and the values were used to calculate the total carotenoid content (µg/g FW or DW) following the proper formulas reported in Lichtenthaler [34].

To determine total phenolic compounds, flavonoids, and anthocyanins content, as well as flowers radical scavenging activity (DPPH and FRAP assays), 200 mg of fresh flowers and 50 mg of dried flowers were extracted in 2 mL of 70% (*v*/*v*) methanol solution for 30 min at 4 °C. At the end of the incubation, the samples were centrifuged at the maximum speed for 10 min, and the supernatants were used for the above mentioned analysis. Total phenolics were determined with Folin-Ciocalteu reagent according to Singleton and Rossi [35], and the results were expressed as mg gallic acid equivalent (GAE) per g FW or DW. The total flavonoids content was determined as reported in Kim et al. [36]. The absorbance was read at 510 nm and the concentration was expressed as mg of (+)-catechin equivalents (CE) per g of FW or DW.

The total monomeric anthocyanin content in the extracts was determined through the pH differential method as described by Lee et al. [37] and Giusti and Wrolstad [38]. Samples were diluted in aqueous buffer at pH 1 (0.025 M potassium chloride buffer) and pH 4.5 (0.4 M sodium acetate buffer) and the absorbance was read at 510 and 700 nm. The monomeric anthocyanin pigment concentration was calculated according to the formula reported in Giusti and Wrolstad [38], and the results were expressed in µg of cyanidin-3-O-glucoside (C3G) per gram of FW or DW.

The determination of the 2,2-diphenyl-1-picrylhydrazyl radical (DPPH) scavenging activity was determined according to Brand-Williams et al. [39]. The results of the DPPH free radical scavenging activity were expressed as mg of Trolox equivalent antioxidant capacity (TEAC) for weight of samples. The Ferric ion Reducing Antioxidant Power (FRAP) antioxidant assay was performed according to Szôllôsi et al. [40], and data were reported as µmol$FeSO_4$/g FW or DW.

The total soluble sugars were spectrophotometrically estimated as reported in Das et al. [41], with some modifications. In brief, 0.8 mL of 0.2% (*w*/*v*) anthrone solution were added to 0.2 mL sample, and the absorbance was read at 620 nm after 12 min of incubation at 90 °C. Data were reported as mg glucose per g FW or DW.

All the data presented are the mean of three independent replicates.

2.4. Volatilomes

Fresh and dried flowers were analyzed for their volatile compounds using the Head Space Solide Phase Microextraction (HS-SPME) following the method previously reported by Najar [42]. Briefly, 1 g of each sample was placed in a 25-mL Erlenmeyer flask for 30 min at room temperature (equilibration time). By the ending time, the fiber (100 µm polydimethylsiloxanes (PDMS, Supelco Ltd., Bellefonte, PA, USA) was exposed in the headspace phase of the samples for 5 min and then transferred to the injector of an Agilent 7890B Gas Chromatograph (Agilent Technologies Inc., Santa Clara, CA, USA) equipped with an Agilent HP-5MS (Agilent Technologies Inc., Santa Clara, CA, USA) capillary column (30 m $\times$ 0.25 mm; coating thickness 0.25 µm) and an Agilent 5977B single quadrupole mass detector (Agilent Technologies Inc., Santa Clara, CA, USA). The analytical conditions of the used GC-MS were already reported [42].

2.5. Statistical Analysis

The biochemical data were statistically analyzed by one-way analysis of variance (ANOVA) with software Past3, version 3.15., using either Tukey Honestly Significant Difference (HSD) or the Mann–Whitney test according to the variance homogeneity (Levene test), with a cut-off significance of $p < 0.05$ (letters). The linear correlation between the antioxidant constituents and antioxidant scavenging activity (DPPH and FRAP assay) was determined using Microsoft Excel ® 2013 (Microsoft Corporation, Redmond, WA, USA).

The one-way ANOVA and the bivariate correlation tests (IBM SPSS software, for Windows, Version 25.0. Armonk, NY, USA) were used first to examine the difference between the major compounds and then to reveal if any relationship occurred between these compounds and the enhancement of the temperature. A correlation matrix was used for the measurement of eigenvalues and eigenvectors in PCA, a multivariate analysis performed on volatile compounds (omitted those present in a % lesser than 0.5%). The PCA plot was performed choosing the two highest principal components (PCs), analysis whose objective is to reduce the dimensionality of the multivariate data of the matrix with preserving most of the variance [43]. The two-way hierarchical cluster analysis (HCA) was performed using Ward's method with squared Euclidian distances as a measure of similarity. These two analyses were conducted by the JMP software package 13.0.0 (SAS Institute, Cary, NC, USA).

3. Results and Discussion

Biochemical Analyses

Fresh flowers are used as control for the concentration of bioactive compounds (Table 1). These quantities are closed to other *Agastache* species, recently described in Najar et al. [2]. Similar composition of polyphenolics was observed recently in *A. rugosa* fresh flowers, while flavonoids and anthocyanins differed due to color dissimilarity with *A. aurantiaca* [44]. Flowers were subjected to different artificial air-drying methods (AD), achieving the same weight loss in different time (65, 12, 5, 3 h) in relation to the applied temperature (30, 50, 60, 70 °C, respectively). The freeze-drying (FD) method is the unique treatment in which the water is removed by high pressure and low temperature, reaching comparable dry weight to the other methods at the end of process (81.3–83.7% water loss)

In post-harvest treatment, color is a remarkable characteristic that describes the quality of a dried product [45]. The intensity of the color depends on the presence of pigments (e.g., carotenoids and anthocyanins), their quality and concentration [46]. The FD system maintained a good amount of these metabolites, as the observed highest value of total carotenoids (Table 1). The maintenance of carotenoids content by FD system was already detected in *Tagetes erecta*, and lutein was even higher in FD than in fresh flowers [47]. The other treatments showed significantly different values, depending on temperature and the length of the drying process. The lower content of carotenoids was detected at 50 °C (AD 50 °C), when the treatment was prolonged for 12 h. Carotenoids in plant tissues are susceptible to oxidation when exposed to light, oxygen, high temperatures, enzymes, moisture, and storage [47]. In this work, the temperature and drying time could be responsible for flowers discoloration and carotenoids depletion, as also reported in various flowers [46,47]. This effect could be ascribed also to the activity of lipoxygenase, a thermostable oxide-reductase enzyme that catalyzes the oxidation of carotene [48].

Table 1. Determination of metabolites in *A. aurantiaca* flowers subjected to different treatments. Data are presented as means ± standard error (SE, $n = 3$). Abbreviations: FW = fresh weight; DW = dry weight; GAE—gallic acid equivalents; CE—± catechin equivalents; C3G—cyanidin-3-O-glucoside equivalents; GLU—glucose. sig.= significant post hoc test at $p < 0.05$.

	Fresh Flowers [1]	Freeze-Drying (FD)	Air Dried (AD) 30 °C	Air Dried (AD) 50 °C	Air Dried (AD) 60 °C	Air Dried (AD) 70 °C
Total Carotenoids µg/g DW	125.93 ± 6.73	222.49 ± 4.88 ^{a}	177.83 ± 2.64 ^{c}	86.77 ± 0.46 ^{e}	187.07 ± 3.68 ^{b}	146.05 ± 1.8 ^{d}
Total Anthocyanins µg C3G/g DW	55.80 ± 2.51	366.65 ± 28.32 ^{a}	185.31 ± 42.2 ^{b}	360.96 ± 58.65 ^{ab}	422.56 ± 15.91 ^{a}	245.7 ± 19.74 ^{b}
Total phenolics (TPC) mg GAE/g DW	3.33 ± 0.12	32.58 ± 2.01 ^{ab}	34.26 ± 1.38 ^{a}	24.60 ± 1.40 ^{b}	25.84 ± 1.90 ^{b}	26.07 ± 1.38 ^{b}
Total flavonoids (TFC) mg CE/g DW	1.768 ± 0.03	29.77 ± 0.58 ^{a}	30.93 ± 0.56 ^{a}	22.66 ± 0.67 ^{b}	23.35 ± 0.81 ^{b}	22.62 ± 0.21 ^{b}
Total soluble sugars (TSS) mg GLU/g DW	26.12 ± 0.37	358.91 ± 5.55 ^{c}	212.12 ± 2.96 ^{d}	512.18 ± 2.68 ^{a}	461.31 ± 4.11 ^{b}	341.99 ± 4.77 ^{c}
FRAP activity µmolFeSO$_4$/g DW	19.33 ± 0.07	403.4 ± 21.78 ^{ab}	458.79 ± 12.88 ^{a}	281.75 ± 17.62 ^{c}	306.84 ± 24.77 ^{c}	318.53 ± 4.63 ^{bc}
DPPH activity mg TEAC/g DW	1.49 ± 0.03	19.4 ± 0.48 ^{b}	23.26 ± 0.52 ^{a}	14.87 ± 0.05 ^{c}	17.88 ± 0.81 ^{b}	17.67 ± 0.01 ^{b}

[1] fresh flower's data are reported as g FW^{-1} instead of g DW^{-1}.

Anthocyanins pigments were also detected in *Agastache* flowers (Table 1). The highest content was detected in AD 60 °C (422.56 µg/g DW), FD (366.65 µg/g DW), and AD 50 °C treated flowers (360.96 µg/g DW). The total content of anthocyanins decreased in the other treatments, probably due to either temperature (70 °C) or length of the drying process (30 °C, 65 h), which affected their degradation. The presence of some enzymes, such as polyphenol oxidase (PPO), could also promote the oxidation of these metabolites. PPO is relatively thermolabile, and temperatures above 50 °C can decrease its activity, especially when high temperature are combined with long drying cycles. In fact, anthocyanins are stable at high temperatures, while PPO is thermolabile and significantly inhibited above 80 °C [49,50].

The content of total phenolics (TPC) and flavonoids (TFC) showed a similar trend. The highest content was quantified in AD 30 °C and FD flowers. TPC was quantified as 32.58 and 34.26 mg GAE/g DW in FD and AD 30 °C, while TFC was 29.77 and 30.93 mg CE/g DW, respectively. With the increase of the temperature, either AD 50, 60, or 70 °C, the TPC and TFC declined to the lowest amounts (Table 1), probably due to the activity of PPO [49,50]. These data confirmed the harmful effect on phenolic compounds of high temperature in AD system, as already observed in other flowers [46,47].

The antioxidant activity was monitored with two different methods: the FRAP assay and the radical scavenger activity (DPPH assay). The DPPH scavenger activity of fresh samples is comparable with the values obtained from the other *Agastache* flowers already published [2]. The FRAP assay detected in dried flowers showed the highest activity in the AD 30 °C samples (458.79 µmol FeSO$_4$/g DW) followed by FD flowers (403.45 µmol FeSO$_4$/g DW), then lower values were reported for the other temperatures. The second method used (DPPH assay), confirmed these results, and the highest value was observed in AD 30 °C flowers (23.26 mg TEAC/g DW) followed by FD samples (19.4 mg TEAC/g DW). At temperature higher than 30 °C, a significant decrease in the general scavenging power occurs, as shown by the loss of natural antioxidants of the dried flowers at 50, 60, and 70 °C.

A linear correlation between antioxidant activity and antioxidant compounds can help to define the contribution of each class of metabolites to flower scavenging activity (Table S1). Regardless of the assay performed, a strong correlation ($R^2 > 0.84$) was observed between antioxidant activity (FRAP or DPPH) and phenolics and flavonoids content. Weaker correlations were detected between antioxidant assays and pigments (carotenoids and anthocyanins), probably also due to their limited amounts.

A relevant characteristic of *Agastache* flowers is the sweetness, due to sugar content [2], the highest macronutrient present in flowers [13]. The amount of total soluble sugars (TSS) detected in fresh flowers (26.12 mg Glu)/g DW) can be attributed to their peculiar percentage of water (Table 1). On the other hand, the highest content of total soluble sugars (TSS) was shown in AD 50 °C, followed by AD 60 °C flowers (512.18 e 461.31 mg GLU/g DW, respectively). FD and AD 70 °C flowers showed a remarkable content of TSS too, while the lowest value for dried flowers is observed at AD 30 °C.

The concentration of sugars in the different treatments might be linked to different factors. Dried flowers at 30 °C showed lower sugar content probably due to the prolonged drying cycle (65 h). This may induce a faster sugars consumption, due to flowers respiration as a consequence of the flower senescence process [51]. At 30 °C, it can be assumed that the senescence of flowers is enhanced, and the ethylene action may affect the respiration rate and therefore the content of soluble sugars [52]. No references are given to support the *Agastache* flower as ethylene sensitive, but the obtained results suggest an involvement of this hormone, to be further investigated. However, at high temperature (50–60–70 °C) the activity of ACC synthetase and the reduced accumulation of ACC oxidase, key enzymes of ethylene biosynthesis, were inhibited, as already demonstrated in cut carnation petals [18,53]. Apart from these observations, little is known about the biochemical and physiological changes that occur during the development and senescence of *A. aurantiaca* flowers. Therefore, it would be appropriate to investigate the nature and behavior of this flower during natural senescence.

4. Volatilome Analyses

The chemical composition of volatiles present in fresh and treated flowers is reported in the Table 2. Overall, fifty-one compounds were identified with a total percentage of identification ranging from 97.5% (in FD samples) to 100% (in fresh flowers).

Oxygenated monoterpenes prevailed in all samples and they amounted for at least 67.1% in AD 70 °C to reach more than 90% in fresh flowers. In the fresh flowers, pulegone was the major compound (75.5%) followed by β-caryophyllene and menthone; the three compounds represented 90.3% of global identified fraction. Pulegone was a common compound among all samples where it dominated in AD 60 °C and AD 70 °C samples, while in the remaining ones was the second main constituent preceded only by menthone. Interesting to note that this latter compound was present in a much smaller amount in the fresh sample (5.2%) and increased exponentially to reach 50% of identified fraction in AD 30 °C. All flowers shared the other three constituents beside pulegone and menthone: limonene (ranged from 1.4% in AD 60 °C to 3.4% in AD 30 °C, respectively), isopulegone (from 1.3% in AD 60 °C to 3.4% in AD 30 °C, respectively). β-caryophyllene, initially the second main constituent in fresh samples, was detected in very low percentage with increasing the drying temperature (not exceeded 2% in AD 60 °C). This was an opposite behavior to that observed in the *Thymus daenensis* Celak, where β-caryophyllene increased enhancing the oven temperature [54]. The bivariate correlation in the cited compounds (Table S2) highlighted the absence of correlation between their amounts and the used temperature except for pulegone, which showed a statistically significant negative linear relationship ($p < 0.05$: greater the amount of pulegone is associated to lower temperature).

Table 2. Chemical composition of the volatile organic compounds (VOCs) from *A. aurantiaca* 'Apricot Sprite' flowers. Headspace–Solid Phase Micro-Extraction (HS-SPME) was performed on Gas Chromatography–Mass Spectrometry (GC–MS) with DB-5 capillary column. Data represent mean values of relative percentage (n = 3, ±SD).

	Compounds	Class	LRI	Fresh	(AD) 30 °C	(AD) 50 °C	(AD) 60 °C	(AD) 70 °C	FD
1	3-Methylcyclohexanone	nt	958	–	0.1 ± 0.01	–	0.2 ± 0.03	–	–
2	Sabinene	mh	974	–	0.1 ± 0.01	–	–	–	–
3	1-Octen-3-ol	nt	980	–	0.2 ± 0.01	–	0.1 ± 0.00	–	–
4	3-Octanone	nt	986	–	0.2 ± 0.01	–	–	–	0.3 ± 0.08
5	6-Methyl-5-heptene-2-one	nt	986	–	–	–	0.2 ± 0.05	–	–
6	β-Myrcene	mh	991	0.6 ± 0.02	0.9 ± 0.14	–	0.1 ± 0.02	0.3 ± 0.07	0.1 ± 0.01
7	Limonene	mh	1030	2.5 ± 0.03	3.4 ± 0.40	1.7 ± 0.04	1.4 ± 0.48	1.8 ± 0.93	2.1 ± 0.48
8	Benzyl alcohol	nt	1036	–	–	–	0.4 ± 0.25	–	–
9	Nonanal	nt	1104	–	–	–	0.1 ± 0.01	0.5 ± 0.16	–
10	p-Menthone	om	1154	5.2 ± 0.06	50.0 ± 0.39	43.3 ± 0.89	28.9 ± 1.27	25.7 ± 0.88	43.4 ± 2.09
11	Isomenthone	om	1164	–	4.8 ± 0.30	4.4 ± 0.11	–	2.9 ± 0.13	3.7 ± 0.13
12	Menthofurane	om	1165	1.7 ± 0.60	–	–	3.3 ± 0.44	–	–
13	Isopulegone	om	1177	2.3 ± 0.29	3.4 ± 0.17	2.7 ± 0.03	1.3 ± 0.78	1.8 ± 0.09	2.5 ± 0.14
14	Verbenone	om	1205	–	0.2 ± 0.02	0.3 ± 0.13	0.3 ± 0.05		0.3 ± 0.08
15	Decanal	nt	1206	–	–	–	0.3 ± 0.07	1.6 ± 0.62	–
16	(−)-trans-Isopiperitenol	om	1210	–	–	–	0.10 ± 0.03	–	–
17	1,4-Dimethyl-4-acetylcyclohexene	nt	1226	–	–	1.3 ± 0.11	3.9 ± 0.97	1.1 ± 0.13	–
18	2-Hydroxycineole	om	1228	–	–	–	0.2 ± 0.03	–	–
19	cis-Pulegone Oxide	om	1230	–	–	0.3 ± 0.03	–	0.2 ± 0.05	–
20	Pulegone	om	1237	75.5 ± 1.85	30.2 ± 0.23	32.6 ± 0.01	34.5 ± 1.66	30.8 ± 0.53	35.8 ± 1.39
21	Piperitone	om	1253	–	0.2 ± 0.00	1.2 ± 0.06	4.5 ± 0.55	2.3 ± 0.16	–
22	2-Hydroxy-3-isopropyl-6-methyl-2-cyclohexen-1-one	nt	1274	–	0.8 ± 0.01	0.7 ± 0.01	0.8 ± 0.06		2.2 ± 0.00
23	1-Cyclohexene-1-carboxaldehyde, 4-hydroxy-2,6,6-trimethyl-3-oxo-	nt	1302	–	–	5.6 ± 0.04	5.7 ± 0.73	2.4 ± 0.42	3.5 ± 0.12
24	2-Hydroxypiperitone	om	1302	–	2.0 ± 0.06	–	–	0.4 ± 0.04	–
25	Undecanal	nt	1307	–	–	–	–	0.2 ± 0.05	–
26	2-Methyl-2-(3-methyl-2-oxobutyl)cyclohexanone	nt	1309	–	0.4 ± 0.02	0.4 ± 0.04	0.6 ± 0.09	–	1.1 ± 0.07
27	Citronellic acid)	nt	1314	–	–	0.3 ± 0.04	0.8 ± 0.06	–	–
28	Piperitenone	om	1340	–	0.2 ± 0.01	0.5 ± 0.03	2.1 ± 0.17	1.1 ± 0.05	–
29	Menthofurolactone	om	1354	–	–	–	–	–	0.1 ± 0.08
30	Eugenol	pp	1357	–	0.3 ± 0.03	0.9 ± 0.04	4.6 ± 0.42	1.3 ± 0.04	
31	1,2-Dimethyl-1-cyclodecene	nt	1360	–	–	–	–	–	0.3 ± 0.04
32	8-Methyl-6-nonenoic acid	nt	1373	–	–	–	0.1 ± 0.04	–	
33	cis-trans-Nepetalactone	om	1377	–	–	–	–	–	0.1 ± 0.07
34	β-Caryophyllene	sh	1419	9.6 ± 1.51	0.9 ± 0.12	1.0 ± 0.08	2.0 ± 0.31	1.8 ± 0.16	0.1 ± 0.06
35	Dihydropseudoionone	nt	1456	–	–	–	–	0.7 ± 0.03	–
36	α-Humulene	sh	1456	1.0 ± 0.32	0.1 ± 0.01	0.3 ± 0.01	0.9 ± 0.11	–	–
37	(E)-β-Famesene	sh	1457	0.6 ± 0.32	–	–	–	–	–
38	2-Hydroxy-4,5-dimethylacetophenone	nt	1476	–	–	0.3 ± 0.01	–	–	–
39	Germacrene D	sh	1481	1.0 ± 0.71	–	–	–	–	–
40	Mintlactone	om	1500	–	0.4 ± 0.04	0.8 ± 0.06	0.8 ± 0.04	0.6 ± 0.03	1.0 ± 0.18
41	Isomintlactone	om	1531	–	0.1 ± 0.01	0.2 ± 0.00	0.1 ± 0.08	-	0.4 ± 0.09
42	Lilial	pp	1534	–	–	–	–	0.7 ± 0.16	–
43	Caryophyllene oxide	os	1581	–	0.10 ± 0.01	–	0.1 ± 0.03	–	0.10.01
44	Hedione	nt	1649	–	0.1 ± 0.01	–	–	4.6 ± 0.59	–
45	(7a-Isopropenyl-4,5-dimethyloctahydroinden-4-yl)methanol	nt	1659	–	–	–	–	0.7 ± 0.16	–
46	Octyl ether	nt	1659	–	–	0.7 ± 0.35	0.7 ± 0.08	6.6 ± 0.17	0.4 ± 0.05
47	α-Hexylcinnamaldehyde	pp	1750	–	–	–	–	1.0 ± 0.10	–
48	Dibutyl adipate	nt	1766	–	–	–	–	1.0 ± 0.13	–
49	Dodecahydro-3a,6,6,9a-tetramethylnaphtho[2,1-β]furan	os	1766	–	–	–	–	0.2 ± 0.03	–
50	2-Hydroxycyclopentadecanone	nt	1846	–	–	–	–	1.8 ± 0.25	–
51	Galaxolide	nt	1851	–	–	–	–	3.2 ± 0.71	–
52	Musk T	os	1989	–	–	–	–	1.0 ± 0.07	–
	Number of identified compounds			**10**	**23**	**21**	**30**	**29**	**19**
	Class of compounds			**Fresh**	**(AD) 30 °C**	**(AD) 50 °C**	**(AD) 60 °C**	**(AD) 70 °C**	**FD**
	Monoterpene hydrocarbons (mh)			3.1 ± 0.05	4.4 ± 0.57	1.7 ± 0.04	1.5 ± 0.36	2.1 ± 0.13	2.2 ± 0.37
	Oxygenated monoterpenes (om)			84.7 ± 2.80	91.5 ± 0.40	88.5 ± 0.81	76.1 ± 1.68	67.1 ± 0.55	87.3 ± 0.50
	Sesquiterpene hydrocarbons (sh)			12.2 ± 2.86	1.0 ± 0.13	1.3 ± 0.10	2.9 ± 0.42	1.8 ± 0.16	0.1 ± 0.01
	Oxygenated sesquiterpenes (os)			–	0.1 ± 0.01	–	0.1 ± 0.00	1.2 ± 0.16	0.1 ± 0.07
	Phenylpropanoids (pp)			–	0.3 ± 0.03	0.9 ± 0.04	4.6 ± 0.42	1.7 ± 0.25	–
	Non-terpene derivatives (nt)			–	1.8 ± 0.04	8.0 ± 0.48	13.9 ± 1.97	24.4 ± 0.54	7.8 ± 0.68
	Total identified			**100 ± 0.00**	**99.1 ± 0.33**	**99.5 ± 0.27**	**99.1 ± 0.62**	**98.3 ± 0.74**	**97.5 ± 0.37**

The fluctuation in the menthone and pulegone percentage may be due to their conversion into other compounds as mentioned by Asekun and coworkers [55]. They attributed the loss of these compounds (such as menthone, pulegone, and 1,8-cineole) in fresh and gently oven-dried samples of *Mentha longifolia* shoots to their vaporization or conversion into other compounds. Although drying, in general, may lead to changes in volatile composition and content, some temperatures allow better preservation of volatile constituents, whereas others may cause significant losses of volatiles.

Both freeze-drying and heat treatment with a temperature more or equal to 50 °C increased the non-terpene amount. Noteworthy is the presence of 4-hydroxy-2,6,6-trimethyl-3-oxo-1-cyclohexene-1-carboxaldehyde in most samples, except for fresh flowers and those treated with a low temperature (30 °C). The amount of this compound was around 5.0% in both AD 50 °C and AD 60 °C then to be halved in AD 70 °C, while in FD samples it amounted to 3.5% of the identified fraction. AD 70 °C flower was also characterized by the presence of good percentage of other non-terpenes such as octyl ether (6.6%) and methyl dihydrojasmonate (4.6%). These samples (AD 70 °C) were distinguished by the presence of galaxolide (3.2%), 2-hydroxycyclopentadecanone (1.8%), α-hexylcinnamaldehyde (1.0%), dibutyl adipate (1.0%), and musk T (1.0%). Even though 70 °C seems to induce the synthesis of these compounds, no correlation was observed between their percentages and the increasing of the temperature. The formation of new chemicals may take place [56] as shown in all treatments by Xing [57] and Díaz-Maroto [58]. This can be attributed to oxidations or to the breakdown of cell wall, or to the hydrolysis of glycosylated forms and thus to the release of several compounds. Concerning the flower volatile emission as a response to high temperature, no report was present in the literature. Only the presence of galaxoide was reported in the EO of Turkish *Helichrysum plicatum* subsp. *polyphyllum* and subsp. *isauricum* [59], while hedione naturally occurred in trace in several flowers [60]. AD 60 °C sample showed the highest number of identified compounds and it was noted by its high amount of eugenol (4.6%), piperitone (4.5%), 1,4-dimethyl-4-acetylcyclohexene (3.9%), mentofurane (3.3%), and piperitenone (2.1%). Only piperitone signaled a statistically positive correlation with the temperature effect ($p < 0.05$) (Table S2. The first two compounds of PCA analyses explained more than 66% of the variance (Figure 1).

PC1 (40.7%) axis was responsible for the segregation of the AD 70 °C flowers from the others due to its high amount in non-terpene compounds such as ester (methyl dihydrojasmonate (12)) and ether (galaxoide (17)). In the left high quadrant (negative PC1, positive PC2), AD 60 °C sample was placed in the borderline, while AD 50 °C almost overlapped PC 2 axis. In the opposite quadrant (negative plot in both axes) were gathered fresh with FD and AD 30 °C flowers; this behavior was especially due to the amount of isopulegone and limonene.

The two-way HCA analysis confirmed what was reported by PCA; in fact two main clusters were observed: C1, a homogeneous cluster formed by only AD 70 °C, and C2 englobed the rest of the samples. C2 is further divided into two subclusters: C2.a, homogeneous and included only fresh flowers, and C2.b, which on its own can be divided into 2 subgroups: one formed by AD 30 °C and the other one with FD and AD 50 °C. (Figure 2)

Only a few works on the spontaneous volatile emission from *Agastache* flowers were present in the literature and especially on the temperature effect and freeze-drying treatments. Both volatile organic compounds of fresh flowers and the EO composition of this studied species (*Agastache aurantiaca* (A. Gray) Lint & Epling) were reported by our team in a recent study [2], as well as only the essential oil composition of some species of this genus were previously investigated [30,61–65]. Drying method, a common technique to preserve the quality of herbs, impact directly in their quality [63]. Even though the eugenol was not present in both fresh and FD samples, this result confirmed what was reported by Ion [64], who underlined no statistical difference in its percentage among fresh and those processed by FD. The same authors concluded that although freeze-drying is one of the most recommended techniques for herb drying, significant changes can occur in the chemical composition of the EO from *Ocimum basilicum* and this result was confirmed by the present study. Of note, was the proximity between FD and AD 50 °C. The same observation was noted in the volatiles of tea flowers [65], where the authors deduced that the volatile from samples treated with 60 °C were close to those observed by the FD and thus recommended the AD method at 60 °C for 180 min. Başer and Buchbauer [66] claimed that monoterpenes vaporized more rapidly by increasing drying temperature, while Mashkani [54] reported that their contents significantly reduced by the increase of the oven temperature. This finding was in disagreement with the results reported herein where a very slight difference in the monoterpene percentage was observed in all samples except for AD 70 °C, where the difference was notable (a decrease of about 27%). Chua and coworkers [67] investigated the EO of *Cassia alta*, and showed the loss in volatile content after FD. They assumed that the loss could be due to the reduction of the pressure of drying chamber, although Antal reported in *Mentha spicata* that the reduction of pressure increased the release of volatile compounds [68]. This statement totally agrees with the results of the present work, where an increase of the number of identified fractions was noted in FD in comparison with fresh flowers.

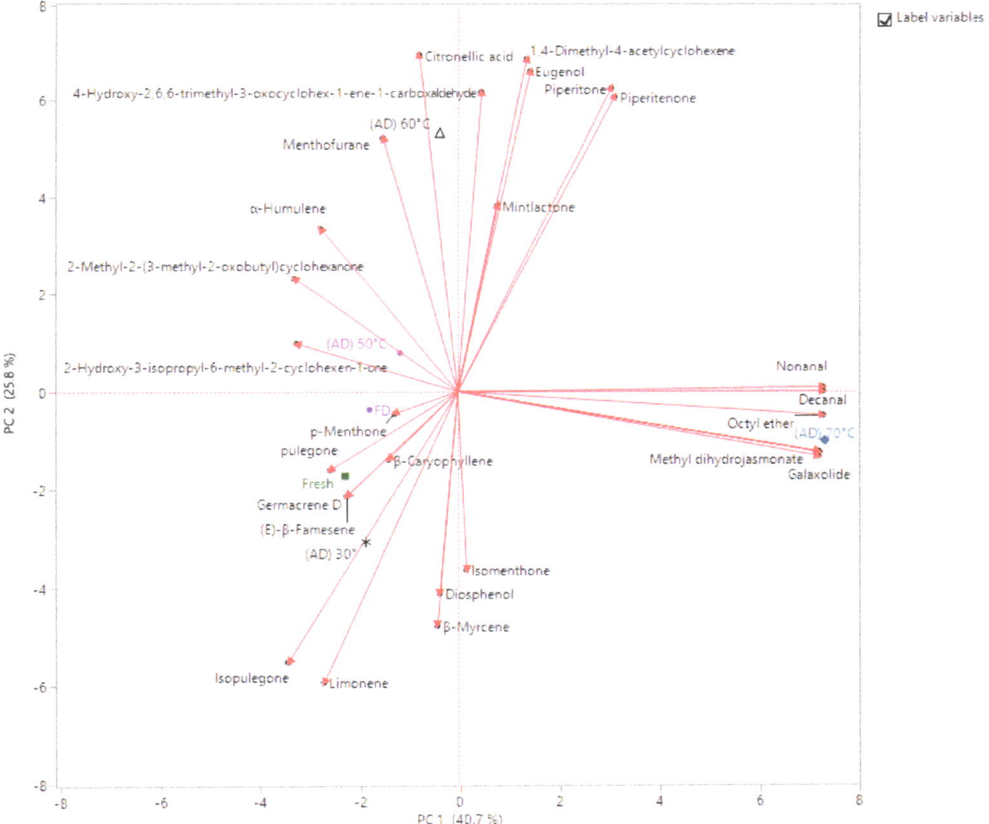

Figure 1. Scatter plot of the principal component analysis (PCA) of the volatile organic compounds (VOCs) of *A. aurantiaca* 'Apricot Sprite' flowers. ■ Fresh Flowers, • Freeze Drying (FD), • (AD) 30°C, • (AD) 50 °C, • (AD) 70 °C.

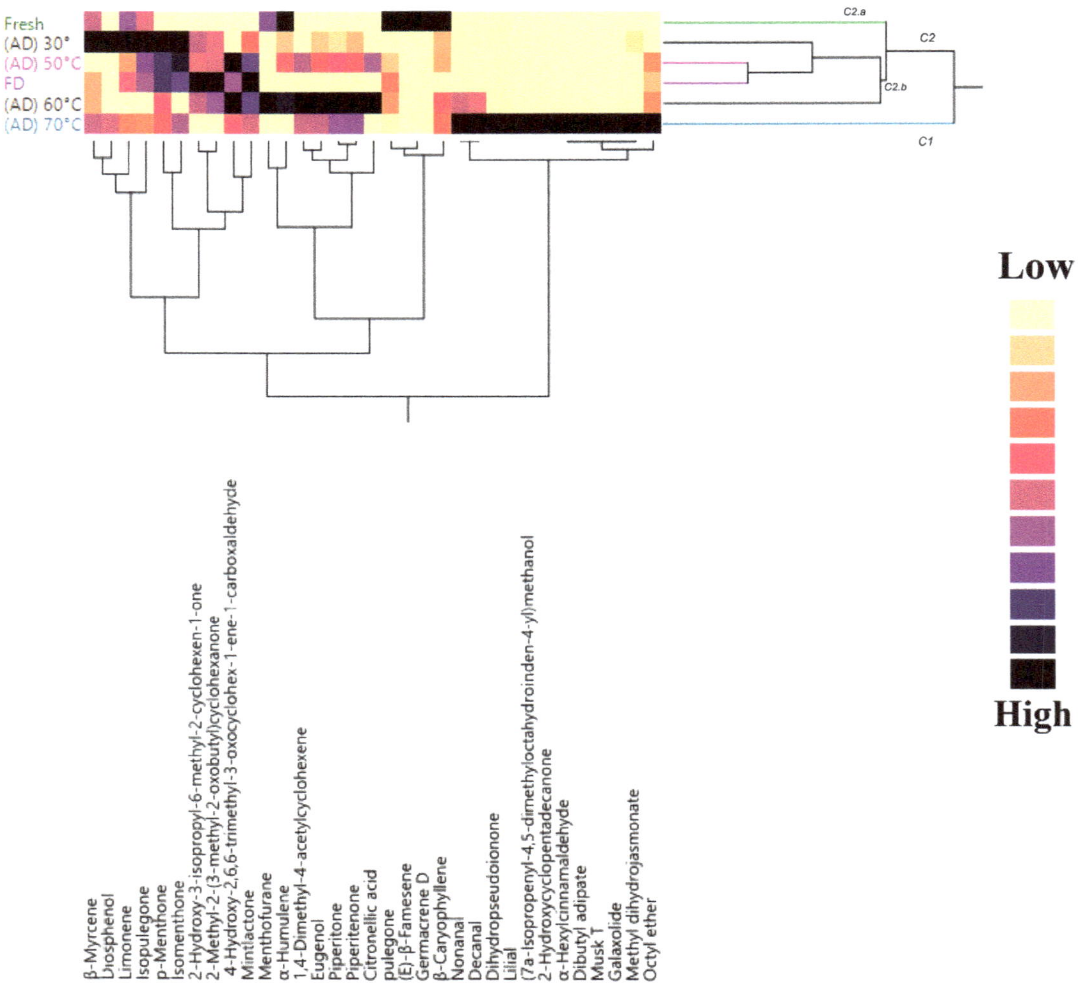

Figure 2. Dendrogram of cluster hierarchical analysis performed on VOCs from *A. aurantiaca* 'Apricot Sprite' flowers.

5. Conclusions

Agastache aurantiaca is an appreciated ornamental plant for their color and aroma, and the flowers are consumed as food, therefore the conservation of their bioactive compounds and aromatic profile is crucial. The presented data showed that the freeze-drying technique is the best solution to prolong the shelf life of these flowers and to maintain high concentration of antioxidant compounds (e.g., phenolic compounds, flavonoids, anthocyanins, carotenoids). The air drying system at 30 °C is a time consuming method with a significant loss of sugars, probably due to the senescence process. The aromatic profile of different treated flowers showed the oxygenated monoterpene compounds as a major class. Pulegone is the main or one of the major constituents of all samples together with *p*-menthone. With the increasing of temperature, pulegone decreased and *p*-menthone increased.

Supplementary Materials: The following are available online at https://www.mdpi.com/article/10.3390/horticulturae7040083/s1. Table S1. Correlation between antioxidants content (total polyphenols, total flavonoids total anthocyanins, total carotenoids) in flowers of *Agastache aurantiaca* and their radical scavenger activity (DPPH-assay and FRAP-assay). Table S2: Bivariate correlation test in the main compounds.

Author Contributions: Conceptualization, L.P. (Laura Pistelli), L.P. (Luisa Pistelli), B.R.; methodology, I.M., B.N., R.D.; software, I.M., B.N.; validation, I.M., B.N., L.P. (Laura Pistelli) and L.P. (Luisa Pistelli); formal analysis, B.N., I.M., R.D.; investigation, B.N., I.M., R.D., G.G.; writing—original draft preparation, B.N., I.M.; writing—review and editing, I.M., B.N., B.R., L.P. (Laura Pistelli), L.P. (Luisa Pistelli); supervision, L.P. (Laura Pistelli), L.P. (Luisa Pistelli); project administration, B.R.; funding acquisition, B.R., L.P. (Laura Pistelli), L.P. (Luisa Pistelli). All authors have read and agreed to the published version of the manuscript.

Funding: "This research was funded by the INTERREG-ALCOTRA UE 2014-2020 Project "ANTEA"–Attività innovative per lo sviluppo della filiera transfrontaliera del fiore edule (n. 1139), grant number: CUP C12F17000080003.

Institutional Review Board Statement: Not applicable.

Informed Consent Statement: Not applicable.

Data Availability Statement: Data is contained within the article.

Acknowledgments: The authors thank Andrea Copetta, Sophie Descamps and Laurent Cambournac for the production and cultivation of plant material.

Conflicts of Interest: The authors declare no conflict of interest.

References

1. Cunningham, E. What nutritional contribution do edible flowers make? *J. Acad. Nutr. Diet.* **2015**, *115*, 856. [CrossRef]
2. Najar, B.; Marchioni, I.; Ruffoni, B.; Copetta, A.; Pistelli, L.; Pistelli, L. Volatilomic analysis of four edible flowers from *Agastache* Genus. *Molecules* **2019**, *24*, 4480. [CrossRef]
3. Li, W.; Song, X.; Hua, Y.; Tao, J.; Zhou, C. Effects of different harvest times on nutritional component of herbaceous peony flower petals. *J. Chem.* **2020**, *2020*. [CrossRef]
4. Drava, G.; Iobbi, V.; Govaerts, R.; Minganti, V.; Copetta, A.; Ruffoni, B.; Bisio, A. Trace elements in edible flowers from Italy: Further insights into health benefits and risks to consumers. *Molecules* **2020**, *25*, 2891. [CrossRef] [PubMed]
5. Marchioni, I.; Najar, B.; Ruffoni, B.; Copetta, A.; Pistelli, L.; Pistelli, L. Bioactive compounds and aroma profile of some Lamiaceae edible flowers. *Plants* **2020**, *9*, 691. [CrossRef] [PubMed]
6. Skrajda-Brdak, M.; Dąbrowski, G.; Konopka, I. Edible flowers, a source of valuable phytonutrients and their pro-healthy effects–A review. *Trends Food Sci. Technol.* **2020**, *103*, 179–199. [CrossRef]
7. Mlcek, J.; Rop, O. Fresh edible flowers of ornamental plants–A new source of nutraceutical foods. *Trends Food Sci. Technol.* **2011**, *22*, 561–569. [CrossRef]
8. Lucarini, M.; Copetta, A.; Durazzo, A.; Gabrielli, P.; Lombardi-Boccia, G.; Lupotto, E.; Santini, A.; Ruffoni, B. A Snapshot on Food Allergies: A Case Study on Edible Flowers. *Sustainability* **2020**, *12*, 8709. [CrossRef]
9. Bourgaud, F.; Gravot, A.; Milesi, S.; Gontier, E. Production of plant secondary metabolites: A historical perspective. *Plant Sci.* **2001**, *161*, 839–851. [CrossRef]
10. Yang, L.; Wen, K.S.; Ruan, X.; Zhao, Y.X.; Wei, F.; Wang, Q. Response of plant secondary metabolites to environmental factors. *Molecules* **2018**, *23*, 762. [CrossRef]
11. Pires, T.C.; Barros, L.; Santos-Buelga, C.; Ferreira, I.C. Edible flowers: Emerging components in the diet. *Trends Food Sci. Technol.* **2019**, *93*, 244–258. [CrossRef]
12. Lu, B.; Li, M.; Yin, R. Phytochemical content, health benefits, and toxicology of common edible flowers: A review (2000–2015). *Crit. Rev. Food Sci. Nutr.* **2016**, *56*, S130–S148. [CrossRef]
13. Fernandes, L.; Casal, S.; Pereira, J.A.; Saraiva, J.A.; Ramalhosa, E. Edible flowers: A review of the nutritional, antioxidant, antimicrobial properties and effects on human health. *J. Food Compos. Anal.* **2017**, *60*, 38–50. [CrossRef]
14. Grzeszczuk, M.; Stefaniak, A.; Meller, E.; Wysocka, G. Mineral composition of some edible flowers. *J. Elem.* **2018**, *23*, 151–162. [CrossRef]
15. Takahashi, J.A.; Rezende, F.A.G.G.; Moura, M.A.F.; Dominguete, L.C.B.; Sande, D. Edible flowers: Bioactive profile and its potential to be used in food development. *Food Res. Int.* **2020**, *129*, 108868. [CrossRef]
16. Khoo, H.E.; Azlan, A.; Tang, S.T.; Lim, S.M. Anthocyanidins and anthocyanins: Colored pigments as food, pharmaceutical ingredients, and the potential health benefits. *Food Nutr. Res.* **2017**, *61*, 1361779. [CrossRef]
17. Zhu, C.; Bai, C.; Sanahuja, G.; Yuan, D.; Farré, G.; Naqvi, S.; Shi, L.; Capell, T.; Christou, P. The regulation of carotenoid pigmentation in flowers. *Arch. Biochem. Biophys.* **2010**, *504*, 132–141. [CrossRef] [PubMed]

18. Hallmann, E. Quantitative and Qualitative Identification of Bioactive Compounds in Edible Flowers of Black and Bristly Locust and Their Antioxidant Activity. *Biomolecules* **2020**, *10*, 1603. [CrossRef]
19. Landi, M.; Ruffoni, B.; Combournac, L.; Guidi, L. Nutraceutical value of edible flowers upon cold storage. *Ital. J. Food Sci.* **2017**, *30*, 1–18. [CrossRef]
20. Fernandes, L.; Casal, S.; Pereira, J.A.; Malheiro, R.; Rodrigues, N.; Saraiva, J.A.; Ramalhosa, E. Borage, calendula, cosmos, Johnny Jump up, and pansy flowers: Volatiles, bioactive compounds, and sensory perception. *Eur. Food Res. Technol.* **2019**, *245*, 593–606. [CrossRef]
21. Marchioni, I.; Pistelli, L.; Ferri, B.; Copetta, A.; Ruffoni, B.; Pistelli, L.; Najar, B. Phytonutritional content and aroma profile changes during postharvest storage of edible flowers. *Front. Plant Sci.* **2020**, *11*, 590968. [CrossRef] [PubMed]
22. Orphanides, A.; Goulas, V.; Gekas, V. Drying technologies: Vehicle to high-quality herbs. *Food Eng. Rev.* **2016**, *8*, 164–180. [CrossRef]
23. Pires, T.C.; Dias, M.I.; Barros, L.; Ferreira, I.C. Nutritional and chemical characterization of edible petals and corresponding infusions: Valorization as new food ingredients. *Food Chem.* **2017**, *220*, 337–343. [CrossRef]
24. Garcìa, L.M.; Ceccanti, C.; Negro, C.; De Bellis, L.; Incrocci, L.; Pardossi, A.; Guidi, L. Effect of Drying Methods on Phenolic Compounds and Antioxidant Activity of *Urtica dioica* L. Leaves. *Horticulturae* **2021**, *7*, 10. [CrossRef]
25. Zielińska, S.; Matkowski, A. Phytochemistry and bioactivity of aromatic and medicinal plants from the genus *Agastache* (Lamiaceae). *Phytochem. Rev.* **2014**, *13*, 391–416. [CrossRef]
26. Duda, M.M.; Varban, D.I.; Muntean, S.; Moldovan, C.; Olar, M. Use of species *Agastache foeniculum* (Pursh) Kuntze. *Hop Med. Plants* **2014**, *21*, 52–54.
27. Husti, A.; Cantor, M.; Buta, E.; Horţ, D. Current trends of using ornamental plants in culinary arts. *ProEnvironment* **2013**, *6*, 52–58.
28. Myadelets, M.A.; Vorobyeva, T.A.; Domrachev, D.V. Composition of the essential oils of some species belonging to genus *Agastache* Clayton ex Gronov (Lamiaceae) cultivated under the conditions of the middle Ural. *Chem. Sustain. Dev.* **2013**, *21*, 397–401.
29. Sanders, R.W. Taxonomy of *Agastache* section Brittonastrum (Lamiaceae-Nepeteae). *Syst. Botany Monogr.* **1987**, *15*, 1–92. [CrossRef]
30. Kovalenko, N.A.; Supichenko, G.N.; Leontiev, V.N.; Shutova, A.G. Composition of essential oil of plants some species of the genus *Agastache* l. Introduced in Belarus. Proceedings of the National Academy of Sciences of Belarus. *Biol. Ser.* **2019**, *64*, 147–155. [CrossRef]
31. Exner, J.; Ulubelen, A.; Mabry, T.J. Chemistry of Agastache. Part II. Flavonoids of *Agastache aurantiaca*. *Rev. Latinoamer. Quim.* **1981**, *12*, 37–38.
32. Saito, N.; Harborne, J.B. Correlations between anthocyanin type, pollinator and flower color in the Labiatae. *Phytochemistry* **1992**, *31*, 3009–3015. [CrossRef]
33. Holloway, P.S.; Matheke, G.E.; Hanscom, J.; Gardiner, A.; Hill, V.; Van Wyhe, E. *Annual Flower Evaluations 2003*; University of Alaska Fairbanks, Agricultural and Forestry Experiment Station, Georgeson Botanical Garden: Sitka, AK, USA, 2003; pp. 1–37.
34. Lichtenthaler, H.K. Chlorophylls and carotenoids: Pigments of photosynthetic biomembranes. *Methods Enzymol.* **1987**, *148*, 350–382.
35. Singleton, V.L.; Rossi, J.A. Colorimetry of total phenolics with phosphomolybdic-phosphotungstic acid reagents. *Am. J. Enol. Viticult.* **1965**, *16*, 144–158.
36. Kim, D.; Jeong, S.; Lee, C. Antioxidant capacity of phenolic phytochemicals from various cultivars of plums. *Food Chem.* **2003**, *81*, 321–326. [CrossRef]
37. Lee, J.; Durst, R.W.; Wrolstad, R.E.; Eisele, T.; Giusti, M.M.; Hach, J.; Hofsommer, H.; Koswig, S.; Krueger, D.A.; Kupina, S.; et al. Determination of total monomeric anthocyanin pigment content of fruit juices, beverages, natural colorants, and wines by the pH differential method: Collaborative study. *J. AOAC Int.* **2005**, *88*, 1269–1278. [CrossRef] [PubMed]
38. Giusti, M.; Wrolstad, R.E. Characterization and Measurement of Anthocyanins by UV-Visible Spectroscopy. *Curr. Protoc. Food Anal. Chem.* **2001**, F1.2.1–F1.2.13. [CrossRef]
39. Brand-Williams, W.; Cuvelier, M.E.; Berset, C.L.W.T. Use of a free radical method to evaluate antioxidant activity. *LWT Food Sci. Technol.* **1995**, *28*, 25–30. [CrossRef]
40. Szôllôsi, R.; Szôllôsi Varga, I. Total antioxidant power in some species of Labiatae (adaptation of FRAP method). *Acta Biol. Szeged* **2002**, *46*, 125–127.
41. Das, B.; Choudhury, B.; Kar, M. Quantitative estimation of changes in biochemical constituents of mahua (*madhuca indica* syn. bassia latifolia) flowers during postharvest storage. *J. Food Process. Preserv.* **2010**, *34*, 831–844. [CrossRef]
42. Najar, B.; Nardi, V.; Cervelli, C.; Pistelli, L. Volatilome analyses of four South African *Helichrysum* spp. grown in Italy. *Nat. Prod. Res.* **2020** [CrossRef]
43. Choi, S.W.; Park, J.H.; Lee, I.B. Process monitoring using a Gaussian mixture model via principal component analysis and discriminant analysis. *Comput. Chem. Eng.* **2004**, *28*, 1377–1387. [CrossRef]
44. Yeo, H.J.; Park, C.H.; Park, Y.E.; Hyeon, H.; Kim, J.K.; Lee, S.Y.; Park, S.U. Metabolic profiling and antioxidant activity during flower development in *Agastache rugosa*. *Physiol. Mol. Biol. Plants* **2021**, *27*, 445–455. [CrossRef]
45. Barani, Y.H.; Zhang, M.; Wang, B.; Devahastin, S. Influences of four pretreatments on anthocyanins content, color and flavor characteristics of hot-air dried rose flower. *Drying Technol.* **2020**, *38*, 1988–1995. [CrossRef]
46. Xu, K.; Zhang, M.; Fang, Z.; Wang, B. Degradation and regulation of edible flower pigments under thermal processing: A review. *CRC Rev. Food Sci.* **2021**, *61*, 1038–1048. [CrossRef]

47. Zhao, L.; Fan, H.; Zhang, M.; Chitrakar, B.; Bhandari, B.; Wang, B. Edible flowers: Review of flower processing and extraction of bioactive compounds by novel technologies. *Food Res. Int.* **2019**, *126*, 108660. [CrossRef] [PubMed]
48. Lončarić, M.; Strelec, I.; Moslavac, T.; Šubarić, D.; Pavić, V.; Molnar, M. Lipoxygenase Inhibition by Plant Extracts. *Biomolecules* **2021**, *11*, 152. [CrossRef]
49. Vamos-Vigyàzò, L. Polyphenol oxidase and peroxidase infruits and vegetables. *CRC Rev. Food Sci.* **1981**, *15*, 49–127. [CrossRef]
50. Severini, C.; Baiano, A.; De Pilli, T.; Romaniello, R.; Derossi, A. Prevention of enzymatic browning in sliced potatoes by blanching in boiling saline solutions. *LWT Food Sci. Technol.* **2003**, *36*, 657–665. [CrossRef]
51. Olley, C.M.; Joyce, D.C.; Irving, D.E. Changes in sugar, protein, respiration, and ethylene in developing and harvested Geraldton waxflower (*Chamelaucium uncinatum*) flowers. *N. Z. J. Crop Hortic.* **1996**, *24*, 143–150. [CrossRef]
52. Rogers, H.J. From models to ornamentals: How is flower senescence regulated? *Plant Mol. Biol.* **2013**, *82*, 563–574. [CrossRef]
53. Yangkhamman, P.; Tanase, K.; Ichimura, K.; Fukai, S. Depression of enzyme activities and gene expression of ACC synthase and ACC oxidase in cut carnation flowers under high-temperature conditions. *Plant Growth Regul.* **2007**, *53*, 155–162. [CrossRef]
54. Mashkani, M.R.D.; Larijani, K.; Mehrafarin, A.; Badi, H.N. Changes in the essential oil content and composition of *Thymus daenensis* Celak. under different drying methods. *Ind. Crops Prod.* **2018**, *112*, 389–395. [CrossRef]
55. Asekun, O.T.; Grierson, D.S.; Afolayan, A.J. Characterization of essential oils from *Helichrysum odoratissimum* using different drying methods. *J. Appl. Sci.* **2007**, *7*, 1005–1008. [CrossRef]
56. Chua, L.Y.W.; Chong, C.H.; Chua, B.L.; Figiel, A. Influence of Drying Methods on the Antibacterial, Antioxidant and Essential Oil Volatile Composition of Herbs: A Review. *Food Bioprocess Technol.* **2019**, *12*, 450–476. [CrossRef]
57. Xing, Y.; Lei, H.; Wang, J.; Wang, Y.; Wang, J.; Xu, H. Effects of different drying methods on the total phenolic, rosmarinic acid and essential oil of purple perilla leaves. J. Essent. *Oil-Bearing Plants* **2017**, *20*, 1594–1606. [CrossRef]
58. Díaz-Maroto, M.C.; Palomo, E.S.; Castro, L.; González Viñas, M.A.; Pérez-Coello, M.S. Changes produced in the aroma compounds and structural integrity of basil (*Ocimum basilicum* L) during drying. *J. Sci. Food Agric.* **2004**, *84*, 2070–2076. [CrossRef]
59. Öztürk, B.; Özek, G.; Özek, T.; Baser, K.H.C. Chemical diversity in volatiles of *Helichrysum plicatum* DC. subspecies in Turkey. *Rec. Nat. Prod.* **2014**, *8*, 373–384.
60. Chapuis, C. The Jubilee of methyl jasmonate and Hedione. *Helv. Chim. Acta* **2012**, *95*, 1479–1511. [CrossRef]
61. Kovalenko, N.N.; Supichenko, G.N.; Akhramovich, T.T.; Shutova, A.G.; Léontiev, V.N. Antibacterial activity of essential oil of *Agastache aurantiaca*. Chem. *Plant Raw Mater.* **2018**, *2*, 63–70. [CrossRef]
62. Yamani, H.; Mantri, N.; Morrison, P.D.; Pang, E. Analysis of the volatile organic compounds from leaves, flower spikes, and nectar of Australian grown *Agastache rugosa*. *BMC Complement. Altern. Med.* **2014**, *14*, 1–6. [CrossRef] [PubMed]
63. Pop, A.; Muste, S.; Păucean, A.; Chiș, M.S.; Man, S.; Salanță, L.; Romina Marc, G.M. A Review of the Drying and Storage Effect on Some Aromatic and Medicinal Plants. Hop Med. *Plants* **2020**, *28*, 142–149.
64. Ion, V.A.; Nicolau, F.; Petre, A.; Bujor, O. Variation of bioactive compounds in organic *Ocimum basilicum* L. during freeze-drying processing. *Sci. Papers Ser. B Hortic.* **2020**, *LXIV*, 397–404.
65. Shi, L.; Gu, Y.; Wu, D.; Wu, X.; Grierson, D.; Tu, Y.; Wu, Y. Hot air drying of tea flowers: Effect of experimental temperatures on drying kinetics, bioactive compounds and quality attributes. *Int. J. Food Sci. Technol.* **2019**, *54*, 526–535. [CrossRef]
66. Başer, K.H.C.; Buchbauer, G. *Handbook of Essential Oils: Science, Technology, and Applications*, 2nd ed.; CRC Press Taylor Francis Group: Boca Raton, FL, USA, 2015. [CrossRef]
67. Wen Chua, L.Y.; Chua, B.L.; Figiel, A.; Chong, C.H.; Wojdyło, A.; Szumny, A.; Lech, K. Characterisation of the convective hot-air drying and vacuum microwave drying of *Cassia alata*: Antioxidant activity, essential oil volatile composition and quality studies. *Molecules* **2019**, *24*, 1625. [CrossRef]
68. Antal, T.; Figiel, A.; Kerekes, B.; Sikolya, L. Effect of drying methods on the quality of the essential oil of spearmint leaves (*Mentha spicata* L.). *Dry. Technol.* **2011**, *29*, 1836–1844. [CrossRef]

Article

Nutraceutical Content and Daily Value Contribution of Sweet Potato Accessions for the European Market

Aline C. Galvao [1], Carlo Nicoletto [1,*], Giampaolo Zanin [1], Pablo F. Vargas [2] and Paolo Sambo [1]

[1] Department of Agronomy, Food, Natural Resources, Animal and Environment–DAFNAE, University of Padova, 35020 Legnaro, Italy; alinecarol.galvao@gmail.com (A.C.G.); paolo.zanin@unipd.it (G.Z.); paolo.sambo@unipd.it (P.S.)

[2] Tropical Root and Starches Center (CERAT), Botucatu, São Paulo State University (UNESP), São Paulo CEP 18.610-034, Brazil; pablo.vargas@unesp.br

* Correspondence: carlo.nicoletto@unipd.it; Tel.: +39-049-8272-826

Abstract: Sweet potatoes (SPs) are considered by the FAO as a primary crop for "traditional agriculture" in the tropics, but in Europe, its consumption is not widespread. However, consumer demand has grown exponentially over the past five years. This study has evaluated the quality and nutrient contents of storage roots of 29 SPs accessions to characterize their role in improving the human diet. Roots were analyzed for nutraceuticals, sugars, and minerals. Results underlined a considerable variability of nutrient content related to color among SPs accessions. The deep-orange-fleshed SPs showed a higher content of β-carotene compared to the light orange- and cream-fleshed ones; 100 g of edible product of HON86 can supply 32.3% of the daily value contribution of vitamin A, followed by the pale orange-fleshed BRA32 and BRA54. The total phenolic content of the purple ecotypes was about two to five times higher than the other genotypes. The calcium content was generally low, whereas, in many accessions, magnesium and phosphorus content reached 20%, or higher of the contribution to the daily value. Such a high variability suggests different use of the different accessions according to their strengths, but might also be used for breeding to improve quality traits of the commercial varieties.

Keywords: *Ipomoea batatas*; nutrients requirement; β-carotene; vitamin A; minerals

Citation: Galvao, A.C.; Nicoletto, C.; Zanin, G.; Vargas, P.F.; Sambo, P. Nutraceutical Content and Daily Value Contribution of Sweet Potato Accessions for the European Market. *Horticulturae* **2021**, *7*, 23. https://doi.org/10.3390/horticulturae7020023

Academic Editor: Lucia Guidi
Received: 31 December 2020
Accepted: 25 January 2021
Published: 30 January 2021

1. Introduction

The sweet potato (*Ipomoea batatas* Lam.) is a tropical herbaceous plant cultivated worldwide that plays a significant role in human and animal nutrition, as well as being a source of starch for the food industry, only in tropical countries [1]. According to the FAO, in a tropical climate sweet potatoes (SPs) are considered a primary crop for "traditional agriculture" and is widely consumed, in particular in Asian and African countries [2]. The worldwide production of SP reached 91.8×10^7 tons in 2019 and is the seventh most cultivated food crop in the world [3].

The main edible parts of SP are its tuberous roots, where significant amounts of sugars [4], as well as vitamins, are stored. SPs are highly adaptive to sub-optimal climatic conditions, and thus they can be cultivated outside of tropical environments; its cultivation in Europe is not widespread, and its consumption is still unusual. However, consumer demand is growing exponentially, thanks to the rising interest in ethnic vegetables mediated by globalization and by nutraceutical research. For these reasons, both imports and consumption are rapidly growing, increasing by 100% over the last five years [5].

The phenological variability of SPs can influence the nutritional contents as reported by several authors [6,7]. The qualitative and nutritional characteristics of the sweet potato are many and concern different categories of nutrients including sugars, fibers, and vitamins, but the parameters that significantly increase their value towards the quality of the diet refer to the content of pigments and polyphenols. As is well known, despite being characterized by a general availability of food and good economic conditions, the European context presents important nutritional deficiencies especially in relation to micronutrients [8–10]. Indeed, red-orange-fleshed SPs rich in β-carotene can be considered as a smart food to fulfill the daily intake of carotenoids as reported by Islam et al. [11]. At the same time, the purple-fleshed varieties may increase the consumer's intake of valuable compounds such as anthocyanins, vitamin C, and oligo-elements that can be used for counteracting food insecurity [11]. Anthocyanins, as well as carotenoids, can be accumulated and stored in plant cells and affect the color of plant tissues. Polyphenols such as carotenoids are involved as scavengers of reactive oxidative species (ROS) in cells. It is widely accepted that anthocyanins might be considered active compounds that reduce the risk of many inflammatory diseases (as observed in vitro) or might regulate diabetes and glycemic levels [12,13]. Although these compounds are present in all varieties of SPs, many studies reported that the highest amount of anthocyanins is contained in purple-fleshed varieties [14–16]. Therefore, orange- and purple-fleshed varieties might be appreciated by health-oriented consumers and hearten the consumption of SPs in Europe.

On the basis of the aforementioned premises, the purpose of the present study was to evaluate the content of nutritional and nutraceutical compounds in roots of 29 accessions of SPs deriving from different countries, Italy included, some of which were cultivated for the first time in Europe.

2. Materials and Methods

2.1. Plant Material

The field experiment was performed at the Experimental Farm of the University of Padova (45°21′ N; 11°58′ E; 6 m a.s.l.). The site has an annual average temperature of 12.3 °C and an annual rainfall of 811 mm. According to the USDA classification, the soil is a silty loam (11% clay, 65% silt, 24% sand) with a pH of 8.2 and 1.65% organic matter.

The propagation material of 29 accessions of SPs (Table 1), used in the experiment, was obtained from the genetic bank of Padova University in cooperation with the Tropical Root and Starches Center (São Paulo State University). On 22 January 2018, SP roots (100–150 g) were potted in a peat-based substrate (Potgrond H Klasmann-Deilmann GmbH, Geeste, Germany) amended with 20% (*v/v*) of perlite (Perlite Italiana, Corsico, Italy) and forced in a greenhouse at a temperature regime of 25/18 °C day/night temperature, under natural light. On 6 June, the cuttings (0.30–0.35 m in length) were collected and transplanted in an open field. Each accession was grown in plots of 7 m^2 with four rows 0.8 m apart and with an in-row spacing of 0.29 m (30 plants per plot). Each plot was replicated three times and arranged in a randomized block design. The soil was fertilized with 80, 70, and 210 kg ha^{-1} of N, P_2O_5, and K_2O, respectively.

Cuttings were manually plugged at a depth of 0.10 m on raised beds and irrigated with about 100 mL of water per cutting.

During the growing period plants were irrigated three times, providing about 30 mm for each irrigation. SPs were then harvested at the commercial size, on 29th September. Harvested roots were cured and stored in the dark at room temperature (17–20 °C and RH of 65–70%) for six weeks before sampling. Sweet potato roots (250–300 g and at least 120 mm long) were randomly taken from each accession. Then stored roots were washed and brushed with tap water to remove any soil residues and dried with tissue paper. Afterward, roots were analyzed for dry matter content by weighing 30 roots per accession and replication before and after drying in a ventilated oven. The other qualitative assessments were performed on the same number of samples as shown below.

2.2. Chemical Reagents

Acetic acid (glacial) and anhydrous sodium carbonate were purchased from Riedel-de Haën (Hanover, Germany). Gallic acid monohydrate was obtained from Fluka (Sigma-Aldrich, Milan, Italy), methanol from VWR Prolabo (Fontenay-sous-Bois, France), and Folin-Ciocalteu's (FC) reagent from Labochimica (Padova, Italy). Chlorogenic acid hemihydrates, maltose, D-(+)-glucose, and D-(−)-fructose were purchased from Aldrich Chemical Company (Sigma-Aldrich); caffeic acid, p-coumaric acid, and ferulic acid from Sigma and methanol from Carlo Erba (Milan, Italy). Deionized water (18 MΩ) was prepared using an ultrapure water purification system (Arium® pro; Sartorius, Muggiò, Italy). All reagents and standards were of analytical and high-performance liquid chromatography (HPLC) grade.

Table 1. List of the sweet potato genetic materials and their main morphological traits.

Genetic Material	Country of Origin	Flesh Color	Skin Color	Roots Shape
Bra1	Brazil	Purple	Dark purple	Elliptic
Bra11	Brazil	Cream	Pink	Round elliptic
Bra13	Brazil	White	Cream	Elliptic
Bra25	Brazil	Purple	Cream	Long oblong
Bra30	Brazil	White	Pink	Long irregular
Bra32	Brazil	Pale orange	Pink	Oblong
Bra33	Brazil	Purple	Cream	Oblong
Bra51	Brazil	White	Cream	Long elliptic
Bra53	Brazil	Purple	White	Oblong
Bra54	Brazil	Intermediate orange	Yellow	Elliptic
Bra66	Brazil	White	White	Long irregular
Bra78	Brazil	Cream	Pink	Long irregular
Bra79	Brazil	Purple	Pink	Obovate
Bra80	Brazil	Purple	Dark purple	Obovate
Hon86	Honduras	Deep orange	Purple-red	Round elliptic
IT41	Italy	Cream	Cream	Long irregular
IT42	Italy	Cream	Pink	Round
IT43	Italy	Pale yellow	Pink	Obovate
IT44	Italy	White	Cream	Elliptic
IT46	Italy	Cream	Pink	Obovate
IT47	Italy	Cream	Cream	Ovate
IT48	Italy	Cream	Cream	Round elliptic
IT49	Italy	Pale yellow	Pink	Round elliptic
IT81	Italy	Cream	Cream	Obovate
IT82	Italy	Cream	Cream	Round elliptic
IT83	Italy	Cream	Cream	Obovate
IT84	Italy	Cream	Cream	Long irregular
USA85	USA	Pale yellow	Cream	Round elliptic
USA45	USA	Purple	Dark purple	Long oblong

2.3. Extraction of Phenols for Analysis

Freeze-dried (Scanvac, Coolsafe model, LaboGene, Denmark) samples (0 5 g) were extracted in methanol (20 mL) with an Ultra Turrax T25 (IKA-Labortechnik, Staufen, Germany) at 13,500 rpm until a uniform consistency was achieved. Samples were filtered (589 filter paper; Whatman, Germany) and appropriate aliquots of extracts were assayed by FC reagent for total phenolic (TP) content and by the ferric reducing antioxidant power (FRAP method) for antioxidant activity. For HPLC analyses, extracts were further filtered by cellulose acetate syringe filters (0.45 μm porosity).

2.4. Determination of TP Content by the FC Assay

The content of TP was determined by the FC assay with gallic acid as a calibration standard, using a UV-1800 spectrophotometer (Shimadzu, Columbia, MD, USA). The FC assay was carried out on 200 μL of SP extract in a 10 mL test tube, followed by the addition of FC reagent (1000 μmL). The mixture was vortexed for 20–30 s and 800 μL of filtered 20% sodium carbonate solution was added within 1–8 min after the FC reagent addition. The mixture was then vortexed again for 20–30 s (time 0). After two hours at room temperature, the absorbance of the colored reaction product was measured at 765 nm by a Shimadzu UV-1800 spectrophotometer (Columbia, MD, USA). The TP content in extracts was calculated from a standard calibration curve obtained with different concentrations of gallic acid, ranging from 0 to 600 μg mL^{-1} (correlation coefficient: $r^2 = 0.9993$). The results are expressed as mg gallic acid equivalent (GAE) kg^{-1} dry weight.

2.5. Determination of Total Antioxidant Activity (TAA) by FRAP

The FRAP reagent was freshly prepared and consisted of 1 mmol L^{-1} of 2,4,6-tripyridyl-2-triazine and 2 mmol L^{-1} of ferric chloride in a 0.25 mol L^{-1} sodium acetate solution (pH 3.6). A methanol extract aliquot (100 μL) was added to the FRAP reagent (1900 μL) and accurately mixed. After four min at room temperature, the absorbance was determined at 593 nm by a Shimadzu UV-1800 spectrophotometer. The calibration was performed with a standard curve (0–1200 μg mL^{-1} Fe^{2+}) (correlation coefficient: $r^2 = 0.9985$), obtained by the addition of freshly prepared ammonium ferrous sulfate. FRAP values were calculated as μg mL^{-1} Fe^{2+} (ferric reducing power) from three determinations and reported as mg kg^{-1} of Fe^{2+} (ferrous ion equivalent) on dry matter.

2.6. Extraction and Determination of Ascorbic Acid

Ascorbic acid was determined by a standard method (ISO 6557/2, 1984). SPs freeze-dried samples (1 g) were homogenized until a uniform consistency was achieved with an Ultra Turrax in 20 mL of meta-phosphoric acid and acetic acid solution. As a colorant reagent, a solution of 2,6-dichlorophenolindophenol was employed. Samples, after adding xylene and 3 min of centrifugation at 4000 rpm, were measured with a Shimadzu UV-1800 spectrophotometer at a wavelength of 500 nm.

2.7. Quantitative Determination of Sugars by HPLC

SPs root freeze-dried samples (0.2 g) were homogenized in demineralized water (20 mL) with an Ultra Turrax T25 at 13,500 rpm until a uniform consistency was achieved. Samples were filtered through filter paper (589; Schleicher), and the extracts were further filtered through cellulose acetate syringe filters (0.45 μm) and analyzed by HPLC. The liquid chromatography apparatus utilized in this analysis was a Jasco X.LC system consisting of a model PU-2080 pump, a model RI-2031 refractive index detector, a model AS-2055 autosampler, and a model CO-2060 column. ChromNAV Chromatography Data System software was used for the analysis of the results. The separation of sugars was achieved on a Hyper-Rez XP Carbohydrate Pb^{++} analytical column (7.7 × 300 mm; Thermo Scientific, Waltham, MA, USA), operating at 80 °C. Isocratic elution was effected using water at a flow rate of 0.6 mL min^{-1}. D-(+)-glucose, D-(–)-fructose, and maltose were quantified by a calibration method. All standards utilized in the experiments were accurately weighed, dissolved in water and the calibration curves were generated with concentrations ranging from 100 to 1000 mg L^{-1} of standards.

2.8. Quantitative Determination of Ions by IC and Organic Nitrogen

For the estimation of anions and cations, each freeze-dried sample (0.2 g) was extracted in water (50 mL) and shaken at 150 rpm for 20 min. Samples were filtered through filter paper (589 Schleicher), and the extracts were further filtered through cellulose acetate syringe filters (0.20 μm) before analysis by ion chromatography (IC). The IC was performed using an ICS-900 Ion Chromatography system (Dionex Corporation) equipped with a dual-piston pump, a model AS-DV autosampler, an isocratic column at room temperature, a DS5 conductivity detector, and an AMMS 300 suppressor (4 mm) for anions and CMMS 300 suppressor (4 mm) for cations. Chromeleon 6.5 Chromatography Management Software was used for system control and data processing. A Dionex Ion-Pac AS23 analytical column (4 mm × 250 mm) and a guard column (4 mm × 50 mm) were used for anion separations, whereas a Dionex IonPac CS12A analytical column (4 mm × 250 mm) and a guard column (4 mm × 50 mm) were used for cation separations. The eluent consisted of 4.5 mM sodium carbonate and 0.8 mM sodium bicarbonate at a flow rate of 1 mL min^{-1} for anions and of 20 mM methanesulphonic acid for cations at the same flow rate. Anions and cations were quantified following a calibration method. Dionex solutions containing seven anions at different concentrations and five cations were taken as standards, and the calibration curves were generated with concentrations ranging from 0.4 mg L^{-1} to 20 mg L^{-1} and from 0.5 mg L^{-1} to 50 mg L^{-1} of standards, respectively.

2.9. *β-Carotene Quantification*

Beta-carotene was measured according to the procedure described by [17,18] with some modifications. For HPLC determination, 0.2 g of freeze-dried material for each cultivar sample was used and carotenoids were extracted with 20 mL of tetrahydrofuran (THF). The mixture was mechanically stirred for 30 s and the extract was filtrated with syringe filters in regenerated cellulose 0.45 µm. The samples were injected into an HPLC model Shimadzu liquid chromatography system (model SCL 10AT VP) equipped with a high-pressure pump (model LC-10AT VP), automatic loop injector (50 µL; model SIL-10AF). Column: Tracer Extrasil ODS2 (250 × 45 mm, 5 µm) using the mobile phase methanol:THF:water (67:27:6, *v/v/v*). The flow-rate of the mobile phase was 1 mL min^{-1} and the absorbance was measured at 470 nm.

2.10. *Starch Determination*

The starch content was determined using the modified AOAC Official Method 996.11 Starch (total) in cereal products and AOAC Official Method 979.10-Starch in Cereals reported by Bulletin 339–2000.

2.11. *Sweet Potato Contribution to Daily Value (SP Contribution to Daily Value)*

The contributions of SPs to the daily recommended intake (daily value, DV) for Vitamin A, Vitamin C, potassium, phosphorus, calcium, and magnesium were calculated by assuming an intake of 100 g fresh product per day. The dietary reference values were set according to the European Food Safety Authority (https://www.efsa.europa.eu/en/interactive-pages/drvs). Values of retinol, vitamin C, potassium, phosphorus, calcium, and magnesium were calculated for adults (>18 years) and children/adolescents (1–17 years). For each value, the corresponding percentage to DV was calculated by the following equation:

DV (%) = nutrient content in 100 g (FW) 100/average daily requirement.

2.12. *Statistical Analysis*

The complete set of data for each variety was analyzed by the PCA procedure using the software Statgraphics Centurion 18.1.06 (Statgraphics Technologies, Inc., The Plains, VA, USA). All qualitative trait data were processed by ANOVA using the software CoStat 6.400 (CoHort Software, CA, USA). The comparison has been realized among all the sweet potato genotypes using a one-way randomized block analysis type. In the case of significant differences, mean values were separated by Tukey HSD test.

3. Results

The highest percentages of dry matter content in the root (>40%) were found in the accessions Bra66, Bra78, Bra80, and IT48, and the lowest (<31%) in Bra30, Bra1, and IT49 (Table 2).

Table 2. Main qualitative composition of the tested sweet potato accessions.

Genetic Materials	Dry Matter (%)	Starch (% dw)	Sucrose g/kg dw	Glucose g/kg dw	Fructose g/kg dw
Bra1	35.9 ^{b}	69.5 ^{c}	76.0 ^{hij}	21.7 ^{def}	16.2 ^{efg}
Bra11	34.7 ^{b}	72.4 ^{b}	88.8 ^{ghi}	19.8 ^{efg}	14.9 ^{efg}
Bra13	32.3 ^{bc}	64.6 ^{c}	106 ^{efg}	8.58 ^{ijk}	6.09 ^{g}
Bra25	35.9 ^{b}	71.6 ^{b}	55.7 ^{jkl}	32.6 ^{bcd}	32.5 ^{bcd}
Bra30	30.7 ^{c}	69.5 ^{c}	68.2 ^{ijk}	46.2 ^{a}	34.1 ^{abcd}
Bra32	32 ^{bc}	71.8 ^{b}	138 ^{abc}	16.7 ^{efg}	14.4 ^{efg}
Bra33	35.9 ^{b}	82.5 ^{ab}	37.5 ^{kl}	27.1 ^{cde}	30.2 ^{cd}
Bra51	29.5 ^{c}	69.6 ^{c}	114 ^{cde}	46.6 ^{a}	43.5 ^{ab}
Bra53	34.3 ^{b}	70.5 ^{bc}	117 ^{cde}	22.1 ^{def}	23.2 ^{def}
Bra54	36.6 ^{b}	77.9 ^{b}	134 ^{bcd}	13.9 ^{ghi}	11.2 ^{fg}
Bra66	41.5 ^{a}	63.6 ^{c}	150 ^{a}	6.72 ^{jk}	7.16 ^{g}
Bra78	41.8 ^{a}	69.8 ^{c}	97.8 ^{fgh}	19.1 ^{efg}	15.1 ^{efg}
Bra79	37.2 ^{b}	71.2 ^{b}	110 ^{cde}	13.7 ^{ghi}	12.0 ^{fg}
Bra80	43.1 ^{a}	74.4 ^{b}	93.2 ^{fgh}	10.5 ^{ghi}	9.50 ^{g}
Hon86	35.9 ^{b}	52.3 ^{d}	95.5 ^{fgh}	43.2 ^{ab}	46.2 ^{a}
IT41	36.6 ^{b}	82.9 ^{ab}	103 ^{efg}	9.30 ^{hij}	7.63 ^{g}
IT42	34.3 ^{b}	72.6 ^{b}	74.0 ^{hij}	38.1 ^{abc}	36.6 ^{abc}
IT43	33.2 ^{bc}	73.4 ^{b}	144 ^{ab}	12.3 ^{ghi}	10.4 ^{g}
IT44	36.8 ^{b}	78.9 ^{b}	109 ^{def}	4.94 ^{k}	3.87 ^{g}
IT46	34.2 ^{b}	74.7 ^{b}	76.3 ^{hij}	15.5 ^{efg}	15.5 ^{efg}
IT47	35.7 ^{b}	91.3 ^{a}	95.9 ^{fgh}	7.38 ^{ijk}	5.16 ^{g}
IT48	40.4 ^{a}	89.2 ^{a}	93.2 ^{fgh}	7.90 ^{ijk}	5.52 ^{g}
IT49	30.4 ^{c}	79.5 ^{b}	92.2 ^{ghi}	36.8 ^{abc}	24.5 ^{cde}
IT81	36.2 ^{b}	72.2 ^{b}	125 ^{bcd}	5.24 ^{k}	3.83 ^{g}
IT82	33.9 ^{bc}	73.4 ^{b}	111 ^{cde}	10.9 ^{ghi}	8.98 ^{g}
IT83	39.5 ^{ab}	76.8 ^{b}	107 ^{def}	7.87 ^{ijk}	5.97 ^{g}
IT84	39.6 ^{ab}	69.9 ^{c}	94.7 ^{fgh}	7.00 ^{jk}	4.43 ^{g}
USA45	35.9 ^{b}	81.4 ^{ab}	36.4 ^{l}	37.0 ^{abc}	35.2 ^{abcd}
USA85	35.9 ^{b}	60.1 ^{cd}	69.6 ^{ijk}	26.4 ^{cde}	31.5 ^{bcd}

Within each parameter, values without common letters significantly differed at $p < 0.05$ according to Tukey's HSD test.

The percentage of starch in SP accessions ranged from 52.3% of Hon86 to 91.3% of IT47. Dividing the accessions according to starch content (<70%, 70–80%, and >80%) into three groups 9 out of the 29 accessions showed low starch content, 15 accessions the intermediate content, and 5 accessions the high content. Considering the origin of the accessions, Brazilian accessions had lower values (71.3%, on average) than the Italian ones (77.5%, on average) (Table 2).

Results on the raw product showed that the sucrose content of Bra66 (150 g kg^{-1} dw) was four-times higher than USA45 (36.4 g kg^{-1} dw) (Table 2). Glucose content in storage roots varied from 4.94 to 46.6 g kg^{-1} dw; Bra51, Bra30, and Hon86 had the highest amount of glucose, with values 10-times higher than the lower one (IT44). Hon86 showed the higher values of fructose (46.2 g kg^{-1} dw), statistically similar to those of Bra51 and IT42 (43.5 and 36.6 g kg^{-1} dw, respectively). Accessions with the lowest content of fructose were found for cream-fleshed SP and white-fleshed SP, such as IT81, IT44, and IT84, with values ranging between 3.83 and 4.43 g kg^{-1} dw.

Figure 1 shows the TP contents and TAA of raw SP roots on a dry weight basis. Content of TP ranged between 870 and 6440 mg GAE kg^{-1} dw. In the purple ecotypes, the TP content was about two to five times higher than other genotypes. The TAA of roots decreased from purple-fleshed > white-fleshed > orange-fleshed > cream-fleshed SP accessions. The higher antioxidant activity was found in purple-fleshed accessions USA45, Bra1, Bra25, and Bra80, with 6700, 4654, 3562, and 2561 mg Fe^{2+}E kg^{-1} dw, respectively.

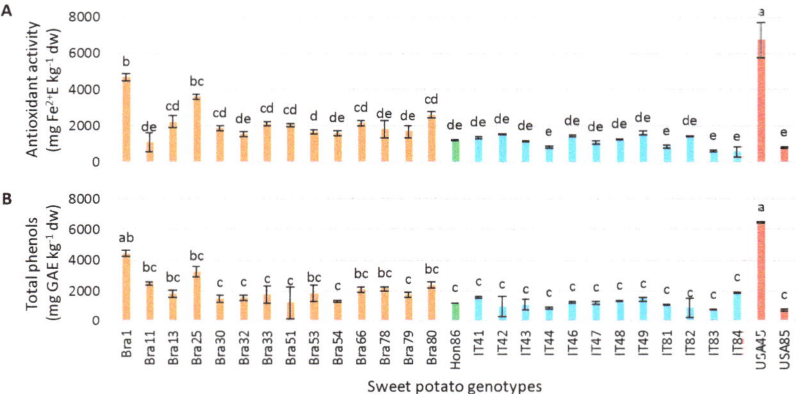

Figure 1. Total phenols content and total antioxidant activity of sweet potato genotypes. Histogram values without common letters significantly differed at $p < 0.05$ according to Tukey's HSD test.

As for the content in mineral elements and vitamins C and A, the values determined in the different accessions have been transformed into the percentage of contribution that a portion of 100 g can provide to the consumer. The reference values used for this processing (SP% contribution to DV) refer to the standards of the European Food Safety Agency (EFSA) and were calculated for both the adult (Table 3) and children (Table 4) consumers.

In Table 3, the higher levels of phosphorus were found in USA45, IT43, IT44, Bra53, and Bra54. One hundred grams of raw SP may provide up to 12.6% contribution to the DV of potassium; the lowest values belong to the colored accessions (purple-, pale-yellow, or orange-fleshed PS), whereas the accessions with pale colors had the highest contribution to the DV. The same pattern was observed for magnesium, and the accessions with the highest contribution to the DV were IT84, IT83, Bra78, and IT44, (39.5%, 31.9%, 28.1%, and 28.0%, respectively). The contribution of the SPs to the DV of calcium ranged between 4.99% and 10.9%. IT84, Bra80, Bra54, Bra79, Hon86, and USA85 accessions had the highest content of calcium. The contribution of the SPs to the DV of vitamin C showed a high variation among the accessions with values ranging from 26.2% to 197%; the cream-fleshed SPs and purple-fleshed SPs were found to have the highest supply of vitamin C.

As for the contribution to the DV of vitamin A, some accessions did not show relevant values. However, Hon86 had the highest values (32.3%). The white-fleshed SPs showed really low values of vitamin A, and in some accessions, the contents were lower than those assessable by the method. The accessions with cream- or light-orange-flesh had the intermediate levels and, in any case, lower than 5% (Table 3). In Table 4, the values of the nutritional parameters have presented the same trends described for the adult consumer, obviously these results are conditioned by a different daily requirement that distinguishes the consumer under the age of 17.

Table 3. Percentage of the contribution to the daily value of minerals and vitamins based on 100 g fresh sweet potato consumption per day for the different accessions according to adults' needs (European Food Safety Agency standards).

Genetic Material	P	K	Na	Cl	Mg	Ca	Vit C	Vit A
Bra1	6.78 ^{ef}	5.44 ^{n}	5.28 ^{a}	3.41 ^{a}	15.2 ^{gh}	7.20 ^{bc}	186 ^{a}	n.d.
Bra11	8.14 ^{ab}	10.4 ^{ef}	2.38 ^{hi}	1.54 ^{ij}	14.4 ^{gh}	6.45 ^{cd}	119 ^{ab}	n.d.
Bra13	6.56 ^{fgh}	6.77 ^{m}	4.93 ^{ab}	3.18 ^{ab}	24.4 ^{de}	6.74 ^{cd}	105 ^{b}	n.d.
Bra25	6.10 ^{hi}	7.25 ^{kl}	3.89 ^{cd}	2.51 ^{cd}	16.2 ^{gh}	7.94 ^{bc}	119 ^{ab}	n.d.
Bra30	3.46 ^{o}	5.94 ^{n}	3.51 ^{ef}	2.26 ^{ef}	18.6 ^{ef}	5.97 ^{de}	n.d.	0.49 ^{c}
Bra32	4.67 ^{lm}	7.40 ^{kl}	5.27 ^{a}	3.40 ^{a}	15.0 ^{gh}	7.74 ^{bc}	167 ^{a}	4.55 ^{b}
Bra33	7.25 ^{cd}	6.99 ^{lm}	3.58 ^{de}	2.31 ^{de}	15.0 ^{gh}	5.97 ^{de}	154 ^{ab}	n.d.
Bra51	5.66 ^{ij}	7.93 ^{ij}	3.77 ^{de}	2.43 ^{de}	17.3 ^{fg}	5.64 ^{de}	126 ^{ab}	0.18 ^{c}
Bra53	8.68 ^{ab}	8.78 ^{gh}	5.14 ^{a}	3.31 ^{a}	17.0 ^{fg}	9.62 ^{bc}	n.d.	n.d.
Bra54	8.49 ^{ab}	6.98 ^{lm}	3.47 ^{ef}	2.24 ^{ef}	27.4 ^{cd}	10.1 ^{ab}	93.0 ^{bc}	3.67 ^{b}
Bra66	7.64 ^{bc}	12.6 ^{bc}	2.65 ^{hi}	1.71 ^{ij}	26.9 ^{cd}	7.84 ^{bc}	106 ^{b}	0.17 ^{c}
Bra78	8.23 ^{ab}	8.18 ^{hi}	4.77 ^{ab}	3.08 ^{ab}	28.1 ^{bc}	7.96 ^{bc}	58.3 ^{d}	n.d.
Bra79	7.62 ^{bc}	12.9 ^{b}	0.88 ^{l}	0.57 ^{l}	19.6 ^{ef}	9.70 ^{bc}	n.d.	n.d.
Bra80	5.91 ^{ij}	7.28 ^{kl}	2.93 ^{gh}	1.89 ^{gh}	26.9 ^{cd}	10.9 ^{a}	190 ^{a}	n.d.
Hon86	7.24 ^{cd}	12.2 ^{c}	1.77 ^{k}	1.14 ^{k}	22.5 ^{de}	8.61 ^{bc}	98.7 ^{bc}	32.3 ^{a}
IT41	7.11 ^{cd}	11.1 ^{d}	2.20 ^{ij}	1.42 ^{jk}	22.5 ^{de}	6.58 ^{cd}	85.8 ^{c}	n.d.
IT42	6.84 ^{de}	11.0 ^{de}	4.47 ^{bc}	2.88 ^{bc}	11.1 ^{h}	5.97 ^{de}	197 ^{a}	0.75 ^{c}
IT43	8.78 ^{ab}	14.3 ^{a}	0.86 ^{l}	0.55 ^{l}	16.4 ^{gh}	6.83 ^{cd}	n.d.	n.d.
IT44	8.37 ^{ab}	8.47 ^{hi}	2.74 ^{gh}	1.77 ^{hi}	28.0 ^{bc}	7.45 ^{bc}	69.2 ^{d}	n.d.
IT46	5.84 ^{ij}	9.97 ^{f}	5.10 ^{a}	3.29 ^{a}	20.0 ^{ef}	5.93 ^{de}	n.d.	n.d.
IT47	4.83 ^{lm}	6.99 ^{lm}	2.54 ^{hi}	1.64 ^{ij}	21.7 ^{de}	6.52 ^{cd}	n.d.	n.d.
IT48	5.11 ^{kl}	7.83 ^{ij}	2.77 ^{gh}	1.79 ^{hi}	23.1 ^{de}	7.53 ^{bc}	92.2 ^{bc}	n.d.
IT49	3.87 ^{no}	6.81 ^{m}	4.03 ^{cd}	2.60 ^{cd}	20.0 ^{ef}	4.99 ^{e}	46.9 ^{d}	1.11 ^{c}
IT81	5.18 ^{jk}	12.6 ^{bc}	2.18 ^{ij}	1.40 ^{jk}	27.9 ^{bc}	7.33 ^{81}	102 ^{b}	n.d.
IT82	6.69 ^{fgh}	12.5 ^{bc}	2.09 ^{jk}	1.35 ^{jk}	22.3 ^{de}	6.41 ^{cd}	103 ^{b}	n.d.
IT83	6.90 ^{de}	9.89 ^{f}	3.35 ^{fg}	2.16 ^{fg}	31.9 ^{ab}	8.45 ^{bc}	72.4 ^{d}	n.d.
IT84	4.38 ^{mn}	5.57 ^{n}	3.30 ^{fg}	2.13 ^{fg}	39.5 ^{a}	9.31 ^{bc}	129 ^{ab}	n.d.
US45	8.99 ^{a}	7.56 ^{jk}	5.02 ^{ab}	3.24 ^{ab}	24.8 ^{de}	6.83 ^{cd}	26.2 ^{e}	n.d.
US85	5.12 ^{kl}	9.19 ^{g}	2.66 ^{hi}	1.71 ^{ij}	20.7 ^{ef}	8.55 ^{bc}	63.5 ^{d}	n.d.

Within each parameter, values without common letters significantly differed at $p < 0.05$ according to Tukey's HSD test. n.d = not detected.

It is interesting to underline that some genotypes can satisfy significant amounts of nutrients such as USA45 for phosphorus (11.2%), IT43 for potassium (27.9%), IT84 for magnesium (46%), and BRA80 for calcium (13%). Most of the accessions considered (>75%) can completely satisfy the need for vitamin C, additionally, HON86 contributes more than 50% of the daily requirement of retinol (vitamin A). The analysis of the PCA and the vector components summarize in a clear way the behavior of the SP accessions in relation to the nutritional potential of a 100 g ration for an adult (Figure 2) and a child (Figure 3). The data of the two categories of consumers are quite similar. However, in both, it is possible to identify three groups of accession able to better satisfy nutritional needs. Particularly, group 1 includes accessions able to supply a significant amount of phosphorus, potassium, and chlorine; for group 2 calcium and magnesium are the most present elements; finally, the third group is characterized by a higher concentration of sodium and vitamin C. Some genotypes have intermediate characteristics, such as IT47 and IT44, which are placed in the center of the figure, and Bra11 which is positioned between groups 1 and 3.

Table 4. Percentage of the contribution to the daily value of minerals and vitamins based on 100 g fresh sweet potato consumption per day for the different accessions according to children needs (European Food Safety Agency standards).

Genetic Material	P	K	Na	Cl	Mg	Ca	Vit C	Vit A
Bra1	8.47 ef	10.6 n	6.21 a	2.57 bc	17.7 fg	8.55 ab	419 a	n.d.
Bra11	10.2 bc	20.1 ef	2.80 hi	2.58 bc	16.8 gh	7.65 bc	268 b	n.d.
Bra13	8.21 gh	13.2 m	5.80 ab	2.02 hi	28.4 bc	8.00 bc	238 b	n.d.
Bra25	7.63 hi	14.1 kl	4.58 cd	1.95 ij	18.9 fg	9.43 ab	267 b	n.d.
Bra30	4.32 o	11.5 n	4.13 de	2.83 ab	21.7 cd	7.09 cd	n.d.	0.87 c
Bra32	5.83 lm	14.4 kl	6.20 a	2.85 ab	17.5 fg	9.19 ab	376 ab	8.11 b
Bra33	9.06 cd	13.6 lm	4.21 de	1.23 m	17.4 fg	7.09 cd	348 ab	n.d.
Bra51	7.07 ij	15.4 ij	4.44 de	2.05 hi	20.2 de	6.70 de	284 b	0.32 c
Bra53	10.9 bc	17.1 gh	6.04 a	2.22 ef	19.8 ef	11.4 ab	n.d.	n.d.
Bra54	10.6 bc	13.6 lm	4.08 de	2.32 ef	32.0 bc	12.0 ab	209 bc	6.53 b
Bra66	9.54 cd	24.5 bc	3.12 hi	2.97 a	31.4 bc	9.31 ab	239 b	0.32 c
Bra78	10.3 bc	15.9 hi	5.61 ab	2.55 bc	32.8 bc	9.45 ab	131 c	n.d.
Bra79	9.52 cd	25.1 b	1.04 l	2.36 ef	22.8 cd	11.5 ab	n.d.	n.d.
Bra80	7.39 ij	14.2 kl	3.44 fg	2.48 cd	31.3 bc	13.0 a	427 a	n.d.
Hon86	9.05 cd	23.7 c	2.08 k	2.27 ef	26.2 bc	10.2 ab	222 b	57.5 a
IT41	8.89 de	21.6 d	2.59 ij	2.26 ef	26.3 bc	7.81 bc	193 bc	n.d.
IT42	8.55 ef	21.4 de	5.26 bc	1.74 kl	12.9 h	7.09 cd	444 a	1.34 c
IT43	11.0 ab	27.9 a	1.01 l	2.73 ab	19.1 fg	8.11 bc	n.d.	n.d.
IT44	10.5 bc	16.5 hi	3.23 gh	1.49 lm	32.6 bc	8.85 ab	155 c	n.d.
IT46	7.30 ij	19.4 f	6.00 a	1.36 m	23.4 cd	7.04 cd	n.d.	n.d.
IT47	6.04 lm	13.6 lm	2.99 hi	2.18 fg	25.4 bc	7.74 bc	n.d.	n.d.
IT48	6.38 kl	15.2 ij	3.26 gh	2.45 de	27.0 bc	8.94 ab	207 bc	n.d.
IT49	4.84 no	13.2 m	4.74 cd	2.18 fg	23.3 cd	5.93 e	105 d	1.97 c
IT81	6.47 jk	24.5 bc	2.56 ij	2.11 gh	32.5 bc	8.70 ab	230 b	n.d.
IT82	8.36 fg	24.4 bc	2.46 jk	2.33 ef	26.0 bc	7.61 bc	233 b	n.d.
IT83	8.63 de	19.2 f	3.94 ef	2.11 gh	37.3 ab	10.0 ab	162 c	n.d.
IT84	5.47 mn	10.8 n	3.89 ef	1.85 jk	46.0 a	11.1 ab	292 b	n.d.
USA45	11.2 a	14.7 jk	5.90 ab	2.29 ef	29.0 bc	8.11 bc	59.0 e	n.d.
USA85	6.40 kl	17.9 g	3.12 hi	1.83 jk	24.2 cd	10.1 ab	142 c	n.d.

Within each parameter, values without common letters significantly differed at $p < 0.05$ according to Tukey's HSD test. n.d = not detected.

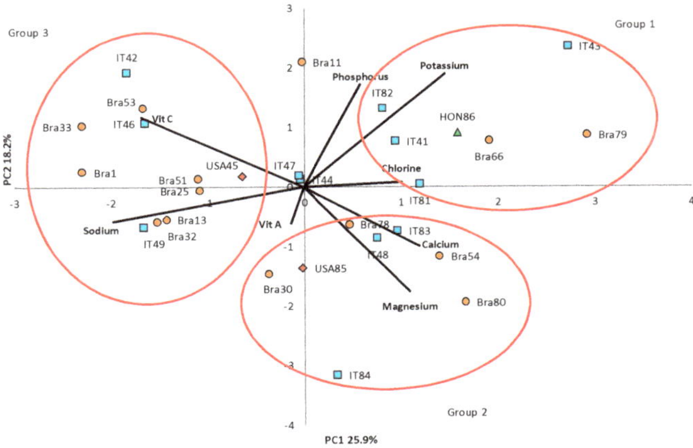

Figure 2. Principal components analysis (PCA) of the sweet potato core collection based on qualitative traits and their contribution to the daily nutritional intake according to adults' needs (European Food Safety Agency standards). Genotypes origin: ▲ Honduras; ◆ United States; ◻ Italy; ● Brazil.

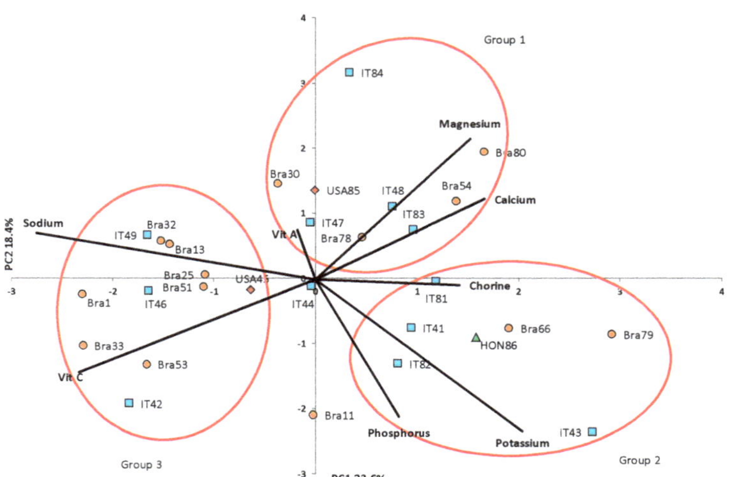

Figure 3. Principal components analysis (PCA) of the sweet potato core collection based on qualitative traits and their contribution to the daily nutritional intake according to children's needs (European Food Safety Agency standards). Genotypes origin: ▲ Honduras; ◆ United States; ■ Italy; ● Brazil.

4. Discussion

The overall nutritional picture that emerged from this work highlights the high potential that SP can offer the consumer in relation to the accession. From a qualitative point of view, the content of starch and soluble sugars represents a relevant aspect considering that the carbohydrate content is directly linked to the sensory aspects that characterize this product.

Roots of SPs are generally stored before the commercialization to increase their sweetness index, so the content of the main metabolic sugars such as glucose, sucrose, and fructose can be influenced either by post-harvest conditions or by the specific physiology of every accession. Indeed, sugar composition is specific to every accession. Sucrose and starch are largely responsible for the post-harvest behavior and the taste of the SP. The cooking procedure can further modify the sugar composition [19]. Conversely, four breeding lines and five commercial varieties of SPs analyzed by Lewthwaite et al. [20], showed that orange-fleshed SPs had a higher content of glucose and fructose compared to the others. Similarly, Agnes et al. [21] found a relationship between sucrose content and color of the flesh: the white-fleshed SP contained less sucrose than the orange-fleshed SP and yellow-fleshed SP. An opposite trend was found for the glucose and fructose contents, as well. Sucrose content is also a physiological signal for the formation of storage roots because there is a trigger of several genes involved in the tissue differentiation. Besides, sucrose is loaded in root cells, and it is converted to hexoses inside amyloplasts and stored as starch. Cervantes-Flores et al. [22] demonstrated that SPs varieties with orange-flesh have lower concentrations of starch. According to them, the concentration of starch and the amount of β-carotene are inversely related, most likely because these two substances compete for the synthesizing sites in the plastids. Our data are in agreement with these findings as the orange genotype Hon86 was characterized by a lower starch content.

The low content of soluble sugars slows the retrogradation of starch, improving the texture after cooking [23]. The genotypes with a lower content of total sugars were Bra33, USA45, and IT48, therefore, they could be used by the food processing industry because the low content of soluble sugars increases the shelf life [24]. As a high amount of sugar can affect the final color and the texture of the food, high levels of simple sugars can interfere with the starch hydration due to the competition with the starch for free water in the flour-water system. During cooking at high temperature, the free sugars interact with proteins and confer brown coloration during the process of caramelization. The dehydration of the carbohydrates at high temperatures provides characteristic color and flavor. Therefore, due to the proportion of sugars, Bra51, Hon86, Bra32, and IT43, could be suitable for frying. It is strategic for the food industry to also consider the ratio of sugars and the starch content of SP roots because these properties can change the chemical-physical properties of the food. The accessions of SP with high starch content, low total sugar, and high dry matter content are suitable for frying, roasting, or baking [21]. Overall, USA45 and Bra33 have interesting characteristics for home cooking. Concerning the dry matter, the values varied between 29.5% and 43.1% and are in line with what was reported by Ellong et al. [25], but higher than that found by Laurie et al. [26] in other genotypes. Therefore, considering the values detected in this experiment, it seems useful to use the classification proposed by Truong et al. [27] which allows to identify all genotypes with a percentage of dry matter higher than 30% as "dry", whereas the BRA51 genotype can be classified as "intermediate".

Cooking methods can influence the nutrient retention of raw SPs, therefore it is essential to use the correct cooking system to maintain as much as possible the nutritional values that the product is able to offer, especially for phenols, vitamin C, or carotenoids. This aspect should be considered for the estimation of the percent of contribution to DV. Several studies have demonstrated that high retention of total carotenoids is reached by using oven drying, with losses ranging from 10 to 4% compared to the raw product. The boiling and the frying processes are characterized by a carotenoid loss from 15% to 10% and from 23% to 15%, respectively [28]. The preservation of carotenoids is essential, as they are linked to protective effects against the mutagenic activity and free radicals [29]. Indeed, regular consumption of food rich in phenols, such as fruits and vegetables, can contribute to the prevention of several diseases, such as cardiac pathology, cancer, and infections [30–32]. Heating can affect the content of the phenolic compounds and antioxidants because the temperature affects the structure of the cells which then release pigments from the cell wall [33]. Nicoletto et al. [19] measured the effect of the cooking process on the qualitative traits of SP and demonstrated an increase in the total phenol and antioxidant capacity content after cooking. For instance, the fried samples had 78% higher antioxidants content than raw samples. Indeed, Padda and Picha [14] compared the phenolic and antioxidant activity and found a similar difference among the 14 SP varieties tested. Our results agree with previous studies, where purple-flashed SPs had a higher amount of phenolic compounds [14,16,34]. The same varieties showed a higher amount of these compounds and the cream- and white-fleshed varieties had a lower content. The genetic variability of the genotypes may be the cause of the variation in phenolic content. The vegetables with red or purple-blue color had a high amount of phenolic and antioxidant compounds, and the levels of some accessions of SPs are comparable to those found in strawberries and blackberries [34].

According to the World Health Organization (WHO), as reported by Stone et al. [35], an insufficient intake of minerals is one of the main risks for human health worldwide. In some countries, the occurrence of this problem is driven by the resource shortages, becoming a leading concern to public health. Significant steps are being taken to address malnutrition: globally, stunting among children under the age of five decreased from 32.6% in 2000 to 22.2% in 2017, whereas there was a slight decrease in underweight women from 11.6% in 2000 to 9.7% in 2016. However, this slow and piecemeal improvement must be pursued further. It should also be considered that the consumption of SP can improve the health of consumers even in developed countries, especially with regard to food with high fiber and a low or medium glycemic index, which can be beneficial for diabetic or insulin-resistant consumers [36]. Moreover, dietary fiber can play an important role in human health revealing a good relationship with the incidence of constipation, obesity, cardiovascular diseases, colon cancer, and diabetes mellitus [37–39]. The diary intake of food with a high antioxidant, phenols, and vitamin contents could be a possible solution for those consuming nutritionally deficient diets.

The WHO and the Europe Food Safety Authority (EFSA) have guidelines for nutrients ingest and the level of necessary intake to satisfy the daily demand for healthy nutrients. For instance, phosphorus plays a role in the metabolism of carbohydrates and fats, making it indispensable for the human body; potassium participates in the osmoregulation of cells, and its absorption by the body is about 90% [35]. According to these aspects, it is important to highlight that some of the accessions considered in this trial can provide a significant amount of phosphorus and potassium such as IT43 and IT81, respectively. Vitamin C, an essential and well-known antioxidant that protects cells against free radical damage, can be supplied in a high amount in SPs. Ten out of the 29 accessions tested in this experiment can fully provide the daily intake by eating a 100 g portion. The assimilation of Vitamin C is high and, during digestion, up to 70–90% is absorbed [40].

Vitamin A is a group of fat-soluble compounds including retinol, retinal, and some esters; it plays a fundamental role in the visual system, immunity, and maintenance of the cells' function. The human body needs a small amount but cannot synthesize it. The demands are more critical in infancy, adolescence, and pregnancy/lactation [41], and vitamin A deficiency is an essential matter in developing countries. Despite the losses due to the cooking process, about 26% [42] of the orange-fleshed SPs showed a higher contribution to the DV. These results agree with those of Burri [43], Huang et al. [42], and Tumwegamire et al. [44]. Therefore, SP can be a considerable a valuable source for a vitamin A deficient population. For example, Burri [43] calculated that one person needs to eat only half a cup of orange-fleshed SP to supply 100% DV of vitamin A. In accordance with what was found by Palumbo et al. [45], the content of vitamin C and vitamin A does not appear to be linked to high concentrations of potassium and calcium, as it is well shown in Figures 2 and 3.

5. Conclusions

According to the obtained data, the consumption of SP could help the intake of phosphorous, potassium, magnesium, calcium, vitamin C, and vitamin A. In particular, orange- and purple-fleshed SPs provide a significant antioxidant and vitamin contribution to the human diet. SP is a staple food in many countries and breeding it to fortify some nutritional values could be the key to enhance the diet value of populations in many regions of the world. In Europe, the growing demand for this product has attracted farmers who are now looking for new opportunities. SP crop production in temperate zones is quite recent, and research is limited. This study has attempted to give guidelines from a nutritional point of view for further research in order to expand both production and consumption of SPs in Europe. Some interesting accessions for consumption were highlighted, in particular USA45 and Bra33, whereas Bra33, USA45, and IT48 showed great traits of their potential use by the food processing industry.

These sweet potato accessions can improve the daily intake of minerals in the diet giving guidelines for deeper research and to expand the production and consumption of sweet potatoes in Europe.

Author Contributions: Conceptualization, A.C.G., C.N., P.S., and P.F.V.; methodology, A.C.G. and C.N.; software, A.C.G. and C.N.; validation, A.C.G. and C.N.; formal analysis, A.C.G., C.N., and G.Z.; investigation, A.C.G. and C.N.; resources, P.S.; data curation, A.C.G. and C.N.; writing—original draft preparation, A.C.G. and C.N.; writing—review and editing, A.C.G., C.N., G.Z., P.S., and P.F.V.; supervision, P.S.; project administration, P.S.; funding acquisition, P.S. All authors have read and agreed to the published version of the manuscript.

Funding: This research received no external funding.

Institutional Review Board Statement: Not applicable.

Informed Consent Statement: Not applicable.

Conflicts of Interest: The authors declare no conflict of interest.

References

1. Mussoline, W.A.; Wilkie, A.C. Feed and fuel: The dual-purpose advantage of an industrial sweetpotato. *J. Sci. Food Agric.* **2017**, *97*, 1567–1575. [CrossRef]
2. Low, J.; Ball, A.; Magezi, S.; Njoku, J.; Mwanga, R.; Andrade, M.; van Mourik, T. Sweet potato development and delivery in sub-Saharan Africa. *Afr. J. Food Agric. Nutr. Dev.* **2017**, *17*, 11955–11972. [CrossRef]
3. FAOSTAT. 2019. Crop Production Statistics. 2018. Available online: http://faostat3.fao.org/home/E (accessed on 30 December 2020).
4. Altemimi, A.B. Extraction and Optimization of Potato Starch and Its Application as a Stabilizer in Yogurt Manufacturing. *Foods* **2018**, *7*, 14. [CrossRef]
5. CBI Market Intelligence. CBI Product Factsheet Fresh Sweet Potatoes in Europe. 2015. Available online: https://it.scribd.com/document/326237170/Product-Factsheet-Europe-Fresh-Sweet-Potatoes-2015 (accessed on 30 December 2020).
6. Neela, S.; Fanta, S.W. Review on nutritional composition of orange-fleshed sweet potato and its role in management of vitamin A deficiency. *Food Sci. Nutr.* **2019**, *7*, 1920–1945. [CrossRef]
7. Cartier, A.; Woods, J.; Sismour, E.; Allen, J.; Ford, E.; Githinji, L.; Xu, Y. Physiochemical, nutritional and antioxidant properties of fourteen Virginia-grown sweet potato varieties. *J. Food Meas. Charact.* **2017**, *11*, 1333–1341. [CrossRef]
8. Bruins, M.J.; Bird, J.K.; Aebischer, C.P.; Eggersdorfer, M. Considerations for Secondary Prevention of Nutritional Deficiencies in High-Risk Groups in High-Income Countries. *Nutrients* **2018**, *10*, 47. [CrossRef]
9. Kehoe, L.; Walton, J.; Flynn, A. Nutritional challenges for older adults in Europe: Current status and future directions. *Proc. Nutr. Soc.* **2019**, *78*, 1–13. [CrossRef]
10. Lockyer, S.; White, A.; Buttriss, J.L. Biofortified crops for tackling micronutrient deficiencies–what impact are these having in developing countries and could they be of relevance within Europe? *Nutr. Bull.* **2018**, *43*, 319–357. [CrossRef]
11. Islam, S.N.; Nusrat, T.; Begum, P.; Ahsan, M. Carotenoids and β-carotene in orange fleshed sweet potato: A possible solution to vitamin A deficiency. *Food Chem.* **2016**, *199*, 628–631. [CrossRef]
12. Zhao, J.-G.; Yan, Q.-Q.; Lu, L.-Z.; Zhang, Y.Q. In vivoantioxidant, hypoglycemic, and anti-tumor activities of anthocyanin extracts from purple sweet potato. *Nutr. Res. Pr.* **2013**, *7*, 359–365. [CrossRef]
13. Salawu, S.O.; Udi, E.; Akindahunsi, A.A.; Boligon, A.A.; Athayde, M.L. Antioxidant potential, phenolic profile and nutrient composition of flesh and peels from Nigerian white and purple skinned sweet potato (Ipomea batatas L.). *Asian J. Plant. Sci.* **2015**, *5*, 14–23.
14. Padda, M.S.; Picha, D.H. Quantification of phenolic acids and antioxidant activity in sweetpotato genotypes. *Sci. Hortic.* **2008**, *119*, 17–20. [CrossRef]
15. Ji, H.; Zhang, H.; Li, H.; Li, Y. Analysis on the Nutrition Composition and Antioxidant Activity of Different Types of Sweet Potato Cultivars. *Food Nutr. Sci.* **2015**, *6*, 161–167. [CrossRef]
16. Teow, C.C.; Truong, V.-D.; McFeeters, R.F.; Thompson, R.L.; Pecota, K.V.; Yencho, G.C. Antioxidant activities, phenolic and β-carotene contents of sweet potato genotypes with varying flesh colours. *Food Chem.* **2007**, *103*, 829–838. [CrossRef]
17. Xu, F.; Yuan, Q.P.; Dong, H.R. Determination of lycopene and & beta-carotene by high-performance liquid chromatography using sudan I as internal standard. *J. Chromatogr. B* **2006**, *838*, 44–49. [CrossRef]
18. Barba, A.O.I.; Hurtado, M.C.; Mata, M.S.; Ruiz, V.F.; De Tejada, M.L.S. Application of a UV–vis detection-HPLC method for a rapid determination of lycopene and β-carotene in vegetables. *Food Chem.* **2006**, *95*, 328–336. [CrossRef]
19. Nicoletto, C.; Vianello, F.; Sambo, P. Effect of different home-cooking methods on textural and nutritional properties of sweet potato genotypes grown in temperate climate conditions. *J. Sci. Food Agric.* **2018**, *98*, 574–581. [CrossRef]
20. Lewthwaite, S.L.; Sutton, K.H.; Triggs, C.M. Free sugar composition of sweetpotato cultivars after storage. *N. Z. J. Crop. Hortic. Sci.* **1997**, *25*, 33–41. [CrossRef]
21. Agnes, N.; Agnes, N.; Yusuf, B.; Judith, N.; Trude, W. Potential use of selected sweetpotato (Ipomea batatas Lam) varieties as defined by chemical and flour pasting characteristics. *Food Nutr. Sci.* **2012**, *3*, 889–896.
22. Cervantes-Flores, J.C.; Sosinski, B.; Pecota, K.V.; Mwanga, R.O.M.; Catignani, G.L.; Truong, V.D.; Watkins, R.H.; Ulmer, M.R.; Yencho, G.C. Identification of quantitative trait loci for dry-matter, starch, and β-carotene content in sweetpotato. *Mol. Breed.* **2011**, *28*, 201–216. [CrossRef]
23. Kohyama, K.; Nishinari, K. Effect of soluble sugars on gelatinization and retrogradation of sweet potato starch. *J. Agric. Food Chem.* **1991**, *39*, 1406–1410. [CrossRef]
24. Chang, S.-M.; Liu, L.-C. Retrogradation of Rice Starches Studied by Differential Scanning Calorimetry and Influence of Sugars, NaCl and Lipids. *J. Food Sci.* **1991**, *56*, 564–566. [CrossRef]
25. Ellong, E.N.; Billard, C.; Adenet, S. Comparison of Physicochemical, Organoleptic and Nutritional Abilities of Eight Sweet Potato (Ipomoea batatas) Varieties. *Food Nutr. Sci.* **2014**, *5*, 196–311. [CrossRef]
26. Laurie, S.M.; van Jaarsveld, P.J.; Faber, M.; Philpott, M.F.; Labuschagne, M.T. Trans-b-carotene, selected mineral content and potential nutritional contribution of 12 sweet potato varieties. *J. Food Compost. Anal.* **2012**, *27*, 151–159. [CrossRef]
27. Truong, V.D.; Hamann, D.D.; Walter, W.M. Relationship between instrumental and sensory parameters of cooked sweetpotato texture. *J. Texture Stud.* **1997**, *28*, 163–185. [CrossRef]
28. Vimala, B.; Nambisan, B.; Hariprakash, B. Retention of carotenoids in orange-fleshed sweet potato during processing. *J. Food Sci. Technol.* **2011**, *48*, 520–524. [CrossRef]

29. Rumbaoa, R.G.O.; Cornago, D.F.; Geronimo, I.M. Phenolic content and antioxidant capacity of Philippine sweet potato (Ipomoea batatas) varieties. *Food Chem.* **2009**, *113*, 1133–1138. [CrossRef]
30. Giglio, R.V.; Patti, A.M.; Cicero, A.F.; Lippi, G.; Rizzo, M.; Toth, P.P.; Banach, M. Polyphenols: Potential Use in the Prevention and Treatment of Cardiovascular Diseases. *Curr. Pharm. Des.* **2018**, *24*, 239–258. [CrossRef]
31. Li, S.; Tan, H.-Y.; Wang, N.; Cheung, F.; Hong, M.; Feng, Y. The Potential and Action Mechanism of Polyphenols in the Treatment of Liver Diseases. *Oxidative Med. Cell. Longev.* **2018**, *2018*, 1–25. [CrossRef]
32. Oak, M.-H.; Auger, C.; Belcastro, E.; Park, S.-H.; Lee, H.-H.; Schini-Kerth, V.B. Potential mechanisms underlying cardiovascular protection by polyphenols: Role of the endothelium. *Free. Radic. Biol. Med.* **2018**, *122*, 161–170. [CrossRef]
33. Huang, Y.-C.; Chang, Y.-H.; Shao, Y.-Y. Effects of genotype and treatment on the antioxidant activity of sweet potato in Taiwan. *Food Chem.* **2006**, *98*, 529–538. [CrossRef]
34. Cevallos-Casals, B.A.; Cisneros-Zevallos, L. Stoichiometric and kinetic studies of phenolic antioxidants from Andean purple corn and red-fleshed sweetpotato. *J. Agric. Food Chem.* **2003**, *51*, 3313–3319. [CrossRef]
35. Stone, M.S.; Martyn, L.; Weaver, C.M. Potassium Intake, Bioavailability, Hypertension, and Glucose Control. *Nutrients* **2016**, *8*, 444. [CrossRef]
36. Allen, C.J.; Corbitt, D.A.; Maloney, P.K.; Butt, S.M.; Truong, V.D. Glycemic index of sweet potato as affected by cooking methods. *Open Nutr. J.* **2012**, *6*, 1–11. [CrossRef]
37. Redondo-Cuenca, A.; Villanueva-Suárez, M.J.; Rodríguez-Sevilla, M.D.; Mateos-Aparicio, I. Chemical composition and dietary fibre of yellow and green commercial soybeans (Glycine max). *Food Chem.* **2007**, *101*, 1216–1222. [CrossRef]
38. Mei, X.; Mu, T.-H.; Han, J.-J. Composition and Physicochemical Properties of Dietary Fiber Extracted from Residues of 10 Varieties of Sweet Potato by a Sieving Method. *J. Agric. Food Chem.* **2010**, *58*, 7305–7310. [CrossRef]
39. Yoshimoto, M.; Yamakawa, O.; Tanoue, H. Potential Chemopreventive Properties and Varietal Difference of Dietary Fiber from Sweetpotato (Ipomoea batatas L.) Root. *Jpn. Agric. Res. Quarterly* **2005**, *39*, 37–43. [CrossRef]
40. World Health Organization. *Global Prevalence of Vitamin a Deficiency in Populations at Risk 1995–2005: WHO Global Database on Vitamin a Deficiency*; WHO Press: Geneva, Switzerland, 2009; Volume 55, ISBN 978 92 4 159801 9.
41. Almeida, L.B.; Penteado, M.V. Carotenoids and pro-vitamin A value of white fleshed Brazilian sweet potatoes (Ipomoea batatas Lam.). *J. Food Compos. Anal.* **1988**, *1*, 341–352. [CrossRef]
42. Huang, A.S.; Tanudjaja, L.; Lum, D. Content of Alpha-, Beta-Carotene, and Dietary Fiber in 18 Sweetpotato Varieties Grown in Hawaii. *J. Food Compos. Anal.* **1999**, *12*, 147–151. [CrossRef]
43. Burri, B.J. Evaluating Sweet Potato as an Intervention Food to Prevent Vitamin A Deficiency. *Compr. Rev. Food Sci. Food Saf.* **2011**, *10*, 118–130. [CrossRef]
44. Tumwegamire, S.; Kapinga, R.; Rubaihayo, P.R.; LaBonte, D.R.; Grüneberg, W.J.; Burgos, G.; Felde, T.Z.; Carpio, R.; Pawelzik, E.; Mwanga, R.O. Evaluation of Dry Matter, Protein, Starch, Sucrose, β-carotene, Iron, Zinc, Calcium, and Magnesium in East African Sweetpotato [Ipomoea batatas (L.) Lam] Germplasm. *HortScience* **2011**, *46*, 348–357. [CrossRef]
45. Palumbo, F.; Galvao, A.C.; Nicoletto, C.; Sambo, P.; Barcaccia, G. Diversity Analysis of Sweet Potato Genetic Resources Using Morphological and Qualitative Traits and Molecular Markers. *Genes* **2019**, *10*, 840. [CrossRef] [PubMed]

horticulturae

Article

Effect of Drying Methods on Phenolic Compounds and Antioxidant Activity of *Urtica dioica* L. Leaves

Leani Martìnez Garcìa [1], Costanza Ceccanti [2,*], Carmine Negro [3], Luigi De Bellis [3], Luca Incrocci [2], Alberto Pardossi [2,4] and Lucia Guidi [2,4]

1 Grupo Tecnico Nacional de Plantas Medicinales de la Asociacion Cubana de Tecnicos Agricolas Agroforestales (ACTAF), Calle 98, Esquina 7ma, Playa, La Habana 11300, Cuba; leanimartinez93@gmail.com
2 Department of Agriculture, Food and Environment, University of Pisa, 56124 Pisa, Italy; luca.incrocci@unipi.it (L.I.); alberto.pardossi@unipi.it (A.P.); lucia.guidi@unipi.it (L.G.)
3 Department of Biological and Environmental Sciences and Technologies, University of Salento, 73100 Lecce, Italy; carmine.negro@unisalento.it (C.N.); luigi.debellis@unisalento.it (L.D.B.)
4 Interdepartmental Research Center Nutrafood "Nutraceuticals and Food for Health", University of Pisa, Via del Borghetto 80, 56124 Pisa, Italy
* Correspondence: c.ceccanti3@studenti.unipi.it

Abstract: Stinging nettle (*Urtica dioica*) is a plant well known in traditional medicine for its many beneficial properties, but the lack of standardization regarding the product to offer to consumers limits its diffusion. To this end, drying appears to be a useful technique to offer a low-cost product that can be stored for long time, but the different drying procedures may give rise to end-products of very different quality as nutraceutical and antioxidant compounds. Nettle leaves have been dehydrated employing freeze-drying (FD), oven-drying (OD) or heat pump drying (HPD) and compared with fresh leaves following water extraction to emulate the use by final consumers. Results indicate that the best dehydration technique is HPD, which apparently gives rise to more than a doubling of total phenols and antioxidant activity in the extract compared to the water extract obtained from fresh leaves but a reduction in the level of ascorbic acid of about 39%. In addition, the content of some phenolic compounds is 10 to over a hundred times higher in the extract after HPD than that obtained from fresh samples. This confirms that the dehydration technique should be tuned in relation to the compounds of greatest interest or value.

Keywords: stinging nettle; freeze-drying; oven-drying; heat pump drying; total phenolic compounds; antioxidant activity

Citation: Garcìa, L.M.; Ceccanti, C.; Negro, C.; De Bellis, L.; Incrocci, L.; Pardossi, A.; Guidi, L. Effect of Drying Methods on Phenolic Compounds and Antioxidant Activity of *Urtica dioica* L. Leaves. *Horticulturae* **2021**, *7*, 10. https://doi.org/10.3390/horticulturae7010010

Received: 13 December 2020
Accepted: 14 January 2021
Published: 19 January 2021

1. Introduction

Urtica dioica L., commonly named stinging nettle, is an herbaceous perennial plant belonging to the family of *Urticaceae* and native of Eurasia. Its medicinal properties are well known, with a wide historical background of the use of its stems, leaves and roots [1]. Recent reviews by Kregiel et al. [2] and Grauso et al. [3] revealed the isolation by stinging nettle leaves of some important phenolic compounds such as caffeic acid, hydroxybenzoic acid, vanillic acid, coumaric acid, gentisic acid, protocatechuic acid, gallic acid, syringic acid, quinic acid and many caffeic and quinic acid derivatives, quercetin, catechin, pelargonidin and apigenin in both the glycosidic and non-glycosidic forms, performed by other authors. The stinging nettle leaves also revealed the presence of important compounds belonging to the terpenoid and carotenoid classes such as 3-oxo-a-ionol, 3-hydroxy-damascone, 3,5,5-trimethyl-4-(1-oxo-2-buten-1-yl)-3-cyclohexen-1-one, 4-(4-hydroxy-2,6,6-trimethyl-1-cyclohexen-1-yl)-3-buten-2-one, 4-(3-hydroxy-1-butyn-1-yl)-3,5,5-trimethyl-2-cyclohexen-1-ol and 3-hydroxy-5,6-epoxy-b-ionol as well as lutein, violaxanthin, neoxanthin, lycopene and β-carotene. Thanks to the phytochemical composition of stinging nettle leaves, this species has traditionally been used to counteract cardiovascular diseases, especially hypertension, but also for its anti-microbial, anti-inflammatory, antirheumatic and acute diuretic effects [2].

Understanding of the molecular mechanisms underlying the nutraceutical effects may open a new horizon for new therapeutic strategies [1].

Natural products, both pure compounds and standardized plant extracts, offer unlimited opportunities as drug sources due to the unequaled availability of chemical diversity. The leaf extract of stinging nettle was one of the herbal remedies for which experimental and clinical trials have complemented each other [1].

Owing to the growing demand for higher productivity of stinging nettle and, at the same time, for higher product quality and lower operating costs, as well as reduced environmental burden, drying technologies have drawn extensive concern. Indeed, oven-drying (OD), the most traditional method, is widely used nowadays for post-harvest processing and storage of a large variety of plant products. However, its long drying time and the use of high temperatures usually decrease product quality because of the degradation of nutritional and nutraceutical compounds, color and flavor. The OD process also has high energy consumption, low efficiency and low productivity. Conversely, the freeze-drying (FD) method may ensure the preservation of the most thermolabile compounds in raw materials, yielding a dried product with high quality. Some authors have studied the quality of plant material following these drying methods, reporting a higher efficiency of the FD method as compared with OD in terms of retention of phenolic compounds, antioxidant activity and other beneficial properties [4–6]. A new drying technology named heat pump drying (HPD), consisting of a constant airflow passage in a sealed environment, is increasingly being used for plants and food because it is an energy-efficient and economically feasible drying device for the production of high-quality dried foods and biomaterials [7]. In an example of this process, the temperature was kept at or below 35 °C. In the first passage of air in the dryer, the air had a low moisture and a constant temperature; then, the same air was cooled, and at this step, the moisture was represented by the sum of the moisture of the air and of the leaves. Subsequently, the same air was heated again and sent back to the main chamber to meet new *U. dioica* leaves to dry [8]. Some authors observed that HPD was more effective in the preservation of phenolic compounds, volatile compounds, ascorbic acid and antioxidant activity in plant materials or fruits than OD and FD [9–11]. However, little information about HPD is available in the literature.

The duration and the temperature of the drying process are the most important factors affecting the chemical, nutraceutical and organoleptic quality of dehydrated plant products, and the selection of the optimal drying method depends on quality requirements, the characteristics of the raw material and the market price of the final product [6].

Therefore, an evaluation of the main classes of nutraceutical compounds at the time of storage could help consumers and producers to understand the potential of this species to be marketed in the most appropriate way. Given all of the above, the objective of this study was to evaluate the effects of three different drying methods on the content of phenolic compounds and ascorbic acid and the antioxidant activity of fresh and dried leaves of *U. dioica* collected in the wild. Leaves were dried with different methods: freeze-drying (FD), oven-drying (OD) and heat pump drying (HPD).

2. Materials and Methods

2.1. Material Preparation

Samples from the species *U. dioica* L. were collected as wild during July 2020 in the "Il Corniolo" farmhouse, located in Castiglione di Garfagnana, Lucca Province, north of Tuscany, Italy (approximately 400 m. a. s. l). Plants with a similar height and leaf number were randomly sampled with a cut approximately 5 cm above the soil. Samples were placed in damp plastic bags to prevent tissue dehydration and transported to the laboratory within 1–2 h. Once in the laboratory, healthy leaves with uniform color, dimensions and texture were detached from the plants and their petioles were removed. Five pooled samples were prepared with the leaves collected from 20 individual plants; afterwards, sub-samples were analyzed fresh or after dehydration with FD, OD or HPD. Sub-samples of fresh plant material were frozen in liquid nitrogen and stored at −80 °C while dried sub-samples were stored at room temperature until laboratory analyses, which were performed within one month from the harvest.

2.2. Leaf Drying

Leaves were freeze-dried (FD) using a standard unheated chamber of dimensions 215 mm Ø × 300 mm height (Telstar LyoQuest-55, Milan, Italy) at a vacuum pressure of 100 Pa and a final condenser temperature of −55 °C until the plant material reached a constant weight (2 days), determining, in this way, the dry weight (DW) of FD leaves. Oven-drying (OD) was performed at 105 °C using a laboratory electric thermostatic oven (Memmert GmbH + Co. KG Universal Oven UN30, Schwabach, Germany) until the plant material reached a constant weight (2 days), determining the DW of OD leaves. An apparatus manufactured by North West Technology, Cuneo, Italy (model NWT-35), was used for heat pump drying (HPD) at a temperature below 35 °C until the plant material reached 15% moisture (2 days for *U. dioica* leaves). The DW of HPD-treated and fresh leaves was determined using the same thermostatic electric oven used for the OD treatment at 60 °C until constant weight.

2.3. Extraction Preparation for Total Phenolic Content and Antioxidant Activity Assays

Leaf samples (1 g) were homogenized in 4 mL of an 80% (*v/v*) aqueous solution of methanol, sonicated with a sonicator (Digital ultrasonic Cleaner, DU-45, Argo Lab, Modena, Italy) for 30 min and then centrifuged with a laboratory centrifuge (MPW 260R, MWP Med. instruments, Warsaw, Poland) at 10,000× *g* for 15 min at 4 °C. Supernatant was stored at −80 °C until analysis.

2.4. Total Phenolic Content

Total phenolic content was determined according to Dewanto et al. [12] with minor modifications. Extracted samples (10 µL) were added to a solution of 115 µL deionized water and 125 µL Folin–Ciocalteau reagent. Blank solution was utilized with 10 µL distilled water. Samples were stirred for 6 min and then 1.25 mL 7% (*w/v*) Na_2CO_3 was added to stop the reaction. Samples were incubated for 90 min at room temperature and the increase in the absorbance at 760 nm was measured against the blank solution with an Ultrospec 2100 Pro spectrophotometer (GE Healthcare Ltd., Little Chalfont, UK). Gallic acid was used as a standard and the results (even for fresh leaves) were expressed as mg gallic acid equivalents per g DW (mg GAE g^{-1} DW).

2.5. 2,2-Diphenyl-1-Picrylhydrazyl Hydrate (DPPH) Free Radical Scavenging Assay

The 2,2-diphenyl-1-picrylhydrazyl hydrate (DPPH) free radical scavenging capacity of each sample was determined according to Brand-Williams et al. [13]. Briefly, 2–10 µL of extract was added to a 3.12×10^{-5} M solution of DPPH in methanol:water 80:20 (*v/v*) to a final volume of 1 mL. The change in absorbance at 515 nm was measured after 30 min of incubation. The antioxidant activity based on the DPPH free radical scavenging ability of the extract (even for fresh leaves) was expressed as mg of Trolox equivalent antioxidant capacity (TEAC) per g DW.

2.6. Ferric Reducing Antioxidant Power (FRAP) Assay

The ferric reducing ability was measured following the method described by Benzie and Strain [14] with some modifications. The ferric reducing antioxidant power (FRAP) reagent was prepared mixing 10 mM 2,4,6-tris(2-pyridyl)-1,3,5-triazine (TPTZ) in 40 mM HCl solution with an identical volume of 20 mM $FeCl_3$ and with ten volumes of 0.3 M acetate buffer (pH 3.6). The mixture was warmed at 37 °C for five minutes prior to the analysis; then, 1 mL of FRAP reagent was mixed with 25–50 µL of sample extract and with distilled water at 37 °C to a final volume of 2 mL. Then, the mixture was incubated at 37 °C for 30 min and absorbance was read at 593 nm. All results were expressed in mg Trolox equivalents per g DW (mg TEAC g^{-1} DW).

2.7. Extracion of Phenolic Compounds

The extraction of phenolic compounds was performed according to Zhou et al. [15] at a ratio of 1:20 (*w/v*). Water extraction was preferred because it has no negative impact on health and the environment. Briefly, the samples were finely ground in a mortar and then extracted with water at 100 °C at reflux for 20 min; they were then centrifuged for 20 min at 5000× *g*, and the supernatant was filtered at 0.45 µm with a polytetrafluoroethylene filter before HPLC analysis.

2.8. Phenolic Compounds' Characterization

Phenolic characterization and quantification on leaf extracts were carried out using an Agilent 1200 High Performance Liquid Chromatography (HPLC) System (Agilent Technologies, Palo Alto, CA, USA) equipped with a standard autosampler and an Agilent Zorbax Extend-C18 analytical column (5 × 2.1 cm, 1.8 µm), as reported by Negro et al. [16]. The HPLC system was coupled to an Agilent diode-array detector (wavelength 280 and 365 nm) and an Agilent 6320 TOF mass spectrometer. Detection was carried out within a mass range of 50−1700 *m/z*. Accurate mass measurements of each peak from the total ion chromatograms (TICs) were obtained by means of an ISO Pump (Agilent G1310B) using a dual nebulizer electrospray ionization source that introduces a low flow (20 µL min^{-1}) of a calibration solution containing the internal reference masses at *m/z* 112.9856, 301.9981, 601.9790 and 1033.9881, in negative ion mode. The compounds were quantified using calibration curves of authentic standards (caffeic acid and rutin). Standards were solubilized in methanol:water 80:20 (*v/v*) and diluted with water:formic acid 99.9:0.1 (*v/v*) at a final concentration of 0.5–10 µg/mL.

2.9. Total Ascorbic Acid Content

Ascorbic acid content was measured spectrophotometrically using the method described by Kampfenkel et al. [17] with some modifications. Extractions were carried out with the homogenization of 0.3 g fresh material with 1 mL 6% (*v/v*) trichloroacetic acid followed by centrifuging for 10 min at 10,000× *g* at 4 °C. Immediately after extraction, the analysis was performed by adding 50 µL supernatant to 50 µL 10 mM dithiothreitol (DTT) and to 100 µL 0.2 M Na-P buffer (pH 7.4). Samples were stirred and incubated for 15 min at 42 °C in a water bath. Then, 50 µL 0.5% (*w/v*) N-ethylmaleimide (NEM) was added and samples were stirred again. After 1 min of stirring, 250 µL 10% (*v/v*) trichloroacetic, 200 µL 42% (*w/v*) orthophosphoric acid, 200 µL 4% (*w/v*) 2,2′-bipyridine (diluted in 70% (*v/v*) ethanol 70% (*v/v*)) and 100 µL 3% (*w/v*) $FeCl_3$ were added to samples. The increase in absorbance at 525 nm was measured against a blank solution (with 6% (*v/v*) trichloroacetic acid instead of supernatant) after 40 min of incubation at 42 °C in a water bath. All results were expressed as mg ascorbic acid per g DW (mg g^{-1} DW).

2.10. Statistical Analysis

To determine the effect of different drying methods, a one-way ANOVA was performed using GraphPad software (GraphPad, La Jolla, CA, USA). Bartlett's test was used to verify the normality of the data. Mean values ± standard deviation (SD) of 5 replicates for each treatment were compared using the least significant difference (LSD) test at p = 0.05.

3. Results and Discussion

Temperature and duration of treatment are the determining factors for the selection of the most efficient drying method to retain phenolic compounds and ascorbic acid in plant materials [18,19]. Indeed, specific variations of temperature in the different drying methods may protect against the degradation of these components, leading to the maintenance or the enhancement of the product quality of the analyzed plant material.

In this study with *U. dioica* leaves, the total phenolic content and antioxidant activity (as analyzed with both DPPH and FRAP assays) were significantly higher in HPD-dried leaves than in fresh leaves and in those dried with FD and OD (Figures 1 and 2). Then, the FD-dried leaves had higher phenolic content and antioxidant activity than the OD-dried leaves, whilst the lowest values of these quantities were found in fresh material (Figures 1 and 2).

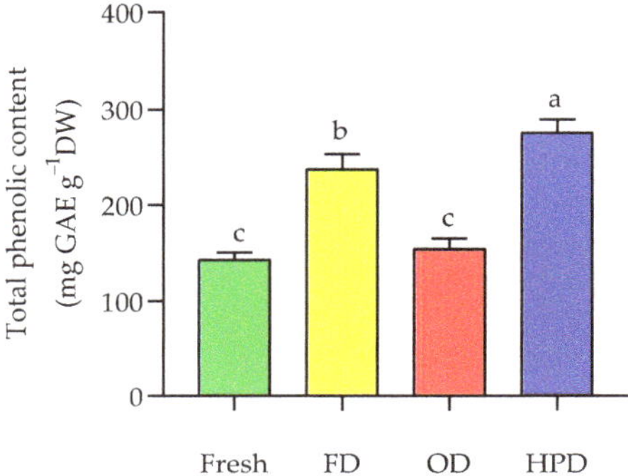

Figure 1. Total phenolic content of fresh or dried leaves of *U. dioica* plants collected in the wild. Leaves were dried with different methods: freeze-drying (FD), oven-drying (OD) and heat pump drying (HPD). Means keyed with a different letter are significantly different for $p = 0.05$ following least significant difference (LSD) test.

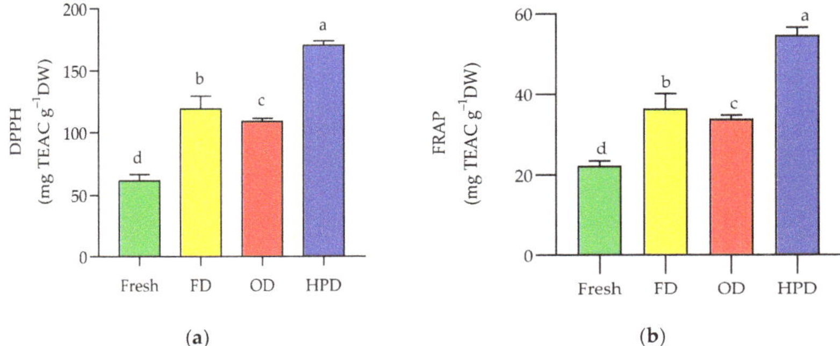

Figure 2. Antioxidant activity of fresh or dried leaves of *U. dioica* plants collected in the wild. Leaves were dried with different methods: freeze-drying (FD), oven-drying (OD) and heat pump drying (HPD). Antioxidant activity was determined with two different assays: 2,2-diphenyl-1-picrylhydrazyl hydrate (DPPH) (**a**) and ferric reducing antioxidant power (FRAP) (**b**). Means keyed with a different letter are significantly different for $p = 0.05$ following LSD test.

The higher total phenolic content of HPD leaves compared to FD or OD leaves could be linked to a more efficient extraction of the insoluble phenolic compounds from the dried or fresh leaves—for instance, phenolic acids or condensed tannins [20–22] linked to cell wall polysaccharides or, more specifically, proteins [23,24]. Indeed, the moderate heat treatment (≤35 °C) used by the HPD may have determined the cleavage of the phenolic sugar glycosidic bonds and the formation of phenolic aglycons, which react better with the Folin–Ciocalteu reagent, thus leading to higher values of the total phenolic content [24]. Conversely, the higher (105 °C) temperature in the OD method probably led to the degradation of some phenolic compounds and a more efficient water extraction of phenolic compounds as well as the cleavage of the phenolic sugar glycosidic bonds.

On the other hand, FD resulted to be an intermediate drying method in terms of maintenance of the phenolic compounds or even an enhancement of their extraction from the leaves. In fact, this method involves the formation of small ice crystals inside the cell and their rapid expulsion from the cell during the freezing [6,23]. The rapid expulsion of the ice crystals allows the maintenance of the cell structures and, therefore, the retention of phenolic compounds and vitamins, and the total expulsion of humidity allows a long preservation of plant samples. These characteristics led to the common use of FD as the extraction pre-treatment of plant materials under analysis [25,26]. In the present study, the content of the analyzed phytochemicals was similar in FD leaves and in fresh leaves. The reason for the lower recovery of phenolic compounds from fresh leaves than that from FD leaves remains to be elucidated in further research.

Nevertheless, the phenol content found in fresh *U. dioica* leaves (Figure 1) is in agreement with previous findings [27,28]. For example, Shonte et al. [28] reported 118.4 mg GAE g^{-1} DW in fresh stinging nettle leaves and 121.5 and 128.7 mg GAE g^{-1} DW in FD and OD leaves, respectively. In contrast, lower values (79 mg GAE g^{-1} DW) were found by Carvalho et al. [29] in wild stinging nettle leaves from the Serra de Estrela region in Portugal and by Hudec et al. [30] in stinging nettle leaves (5.38 mg GAE g^{-1} DW) of cultivated plants in the area of Nitra, in Slovakia. The differences in the phenolic contents found in this study and by other authors may be due to genetic differences between the plants used and growing conditions.

Similar to the total phenolic content results, the values of antioxidant activity resulted higher in the HPD leaves when compared with fresh, OD and FD samples (Figure 2a,b). The higher antioxidant activity in HPD leaves with respect to the other leaves could be due to different reasons: (i) release of bound antioxidant substances (such as phenolic compounds) by disrupting the cell structure and by obtaining a natural extraction for the analysis [31]; (ii) formation of new antioxidants such as Maillard reaction products (melanoidins) resulting from thermal chemical reaction; (iii) suppression of oxidation by thermal inactivation of oxidative enzymes of antioxidants such as polyphenol oxidase [32]. Indeed, a heat treatment at moderate temperature (50 °C for 5 h) inactivated polyphenol oxidases in extracts from aerial parts of *Phyllanthus amarus* [32]. However, the HPLC analysis did not reveal the presence of different phenolic compounds in HPD leaves with respect to the other treatments (Figure 3 and Table 1).

The temperature (105 °C) used in OD leaves apparently resulted in too high of an increase in the release of phytochemicals, even though we observed a slightly higher-level small increase in antioxidant activity in comparison to fresh leaves (Figure 2a,b).

Table 1 shows the HPLC analysis of the stinging nettle leaves under investigation, confirming the total phenolic compounds and antioxidant activities data (Figures 1 and 2). The qualitative analysis mainly shows the presence of some caffeic acid isomers, both free and combined with quinic or malic acid, and some flavonoids such as rutin, isoquercetin and isorhamnetin rutinoside [20–22].

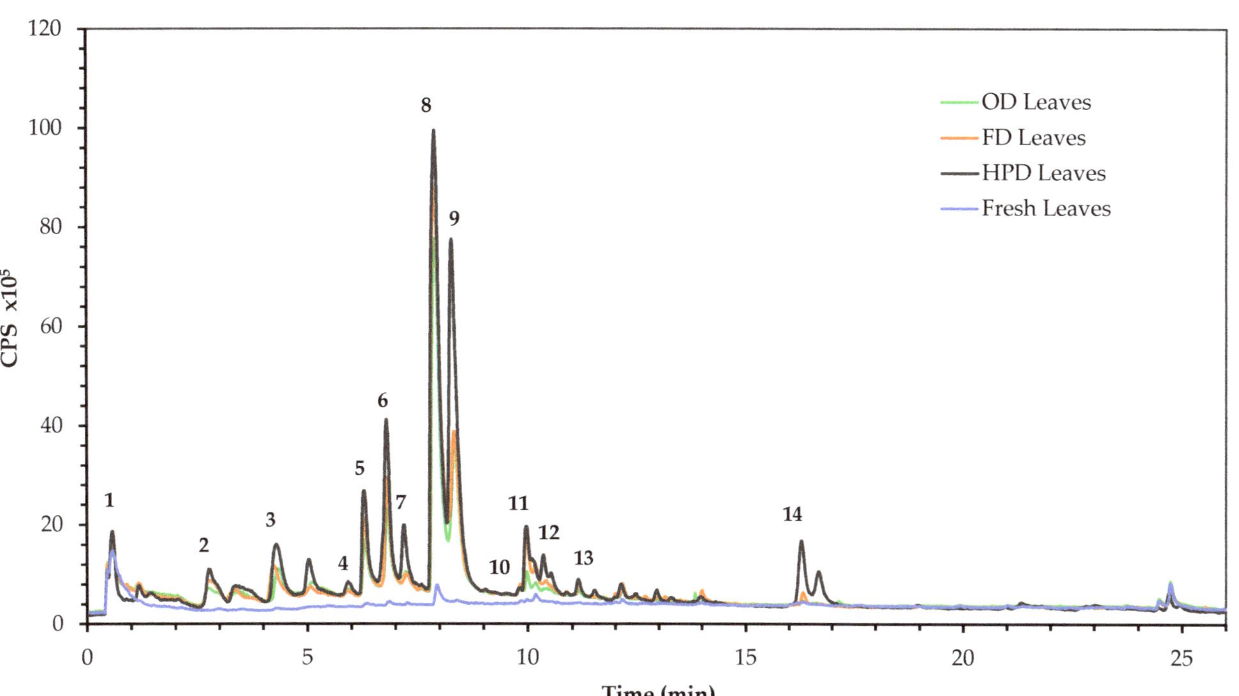

Figure 3. Phenolic profile of fresh or dried leaves of *U. dioica* plants collected in the wild. Leaves were dried with different methods: freeze-drying (FD), oven-drying (OD) and heat pump drying (HPD). The blue line represents the chromatogram of fresh leaves, the orange line the chromatograph of freeze-dried leaves, the green line the chromatogram of the oven-dried leaves and the black line the chromatograph of the heat pump-dried leaves. Numbers refer to the identified phenolic compounds reported in Table 1.

Table 1. High performance liquid chromatography/diode array detector/mass spectrometry-time-of-flight (HPLC/DAD/MS-TOF) tentative identification and quantification of the phenolic compounds extracted from fresh or dried leaves of *U. dioica* plants collected in the wild. Leaves were dried with different methods: freeze-drying (FD), oven-drying (OD) and heat pump drying (HPD).

Peaks	RT (min)	Molecular Ion [M-H]$^-$ (*m/z*)	*m/z* Exp.	*m/z* Calc.	Error (ppm)	Tentative Identification	Quantification (mg/g DW)				References
							Fresh	FD	OD	HPD	
1	0.580	$C_7H_{11}O_6$	191.0574	191.0561	−6.72	Quinic acid *	0.08 ± 0.05 c	0.19 ± 0.05 b	0.18 ± 0.04 b	0.25 ± 0.03 a	[16]
2	2.534	$C_{13}H_{11}O_9$	311.0424	311.0409	−4.86	Unknown	nd	nd	nd	nd	-
3	3.925	$C_{16}H_{17}O_9$	353.0916	353.0878	−10.67	Caffeoylquinic acid 1	0.80 ± 0.05 d	3.88 ± 0.04 b	3.02 ± 0.03 c	7.60 ± 0.02 a	[33,34]
4	5.636	$C_9H_7O_4$	179.0364	179.0350	−8.01	Caffeic acid *	0.15 ± 0.01 b	0.16 ± 0.03 b	0.16 ± 0.03 b	0.40 ± 0.02 a	[33,34]
5	6.177	$C_{16}H_{17}O_9$	353.0907	353.0878	−8.27	Caffeolquinic acid 2	0.10 ± 0.01 d	0.88 ± 0.04 b	0.60 ± 0.03 c	1.92 ± 0.02 a	[33,34]
6	6.754	$C_{16}H_{17}O_9$	353.0885	353.0878	1.98	Caffeoylquinic acid 3	0.75 ± 0.02 d	5.32 ± 0.03 b	4.32 ± 0.03 c	9.60 ± 0.03 a	[33,34]
7	7.255	$C_{16}H_{17}O_9$	353.0914	353.0878	−10.13	Caffeoylquinic acid 4	0.10 ± 0.01 d	0.68 ± 0.02 c	1.04 ± 0.03 b	1.61 ± 0.03 a	[33,34]
8	7.795	$C_{26}H_{23}O_{16}$	591.1059	591.0992	−11.44	Caffeoylmalic acid dimer	0.85 ± 0.02 d	24.72 ± 0.04 b	19.20 ± 0.02 c	35.81 ± 0.03 a	[22,33,35]
9	8.303	$C_{13}H_{11}O_8$	295.0470	295.0450	6.77	Caffeoylmalic acid	0.10 ± 0.01 d	3.02 ± 0.02 c	8.02 ± 0.04 b	13.68 ± 0.05 a	[22,33,35]
10	9.639	$C_{10}H_9O_4$	193.0526	193.0506	−10.11	Ferulic acid *	0.02 ± 0.01 c	0.08 ± 0.01 a	0.05 ± 0.02 b	0.10 ± 0.03 a	[34]
11	9.910	$C_{27}H_{19}O_{12}$	609.1499	609.1461	−6.27	Rutin *	0.30 ± 0.02 d	1.40 ± 0.03 b	0.44 ± 0.02 c	2.25 ± 0.04 a	[22,33,34]
12	10.020	$C_{21}H_{31}O_{16}$	463.0912	463.0882	−6.58	Isoquercetin *	0.21 ± 0.03 d	0.24 ± 0.03 b	0.12 ± 0.04 c	0.84 ± 0.02 a	[33]
13	11.184	$C_{28}H_{31}O_{16}$	623.1639	623.1618	−3.45	Isorhamnetin rutinoside	0.05 ± 0.01 b	0.04 ± 0.02 c	0.02 ± 0.01 d	0.06 ± 0.01 a	[22]
14	16.379	$C_{14}H_{17}O_4$	249.1130	249.1132	1.05	Unknown	nd	nd	nd	nd	-

m/z exp.: *m/z* experimental; *m/z* calc.: *m/z* calculated; Error: difference between the observed mass and the theoretical mass of the compound (ppm); nd: not detected; different letters indicate significant differences for $p = 0.05$ following least significant difference (LSD) test. * Confirmed by the authentic chemical standard.

The most abundant phenolic acid was the caffeoylmalic acid dimer, with values between 19.20 and 35.81 mg g^{-1} DW for OD and HPD samples, respectively. In addition, the isomers of caffeoylquinic acid were more abundant in the HPD sample, with 20.73 mg g^{-1} DW, twice the value obtained from the OD sample (10.76 mg g^{-1} DW), thus confirming that the HPD technique preserves more phenolic compounds. Regarding flavonoids, the most representative one was rutin, whose quantity in the HPD sample reached 2.25 mg g^{-1} DW. The values obtained were coherent with those reported in the literature. In fact, Grevsen et al. [33] reported amounts of caffeoylmalic acid approximately between 7 and 22 mg g^{-1} DW and caffeoylquinic acid values between 3 and 18 mg g^{-1} DW in different accessions of *U. dioica*. Besides, Orčić et al. [34] reported a content of caffeoylquinic acid variable between 1.2 and 28 mg g^{-1} DW, while Vajić et al. [35] showed amounts of caffeoylmalic acid up to 1.03 mg g^{-1} DW in water extract and of 1.64 mg g^{-1} DW in 50% methanol extract. Orčić et al. [34] indicated rutin content between 0.002 and 4.6 mg g^{-1} DW, and Grevsen et al. [33] found contents up to 14.5 mg g^{-1} DW, depending on the harvest time. Virtually all samples showed very high levels after HPD compared to fresh leaves; interestingly, caffeoylmalic acid and its dimer were particularly enriched in HPD samples, up to 42 and 136 times, respectively, while only isorhamnetin rutinoside did not show an increment after dehydration, showing a slight reduction for FD and OD samples and a slight increase for HPD leaves. As discussed previously, this phenomenon deserves further investigation.

While all drying methods increased the total phenolic content and the antioxidant capacity of stinging nettle leaves, the total ascorbic acid content was significantly lower in FD (−36%), OD (−55%) and HPD (−39%) leaves than in fresh leaves (Figure 4). As reported before, FD is usually used as an extraction pre-treatment in plant materials during vitamin and phenolic content analyses. Nevertheless, in the present study, this drying technique resulted in adverse retention of ascorbic acid with a decrease of 36% compared to fresh leaves. However, these findings agree with those reported by Shonte et al. [30]. These authors found a loss of 12% and 22% in FD and OD stinging nettle leaves, respectively, with values of 14.2, 11.8 and 9.9 mg g^{-1} DW in fresh, FD and OD leaves, respectively. Some studies reported that the loss of total ascorbic acid in dried food products can range from 10% to more than 50% depending on drying temperature [28,36]. This loss is mainly due to chemical degradation involving the oxidation of ascorbic acid to dehydroascorbic acid, followed by hydrolysis to 2,3-diketogulonic acid and subsequent polymerization to form other nutritionally inactive products [37].

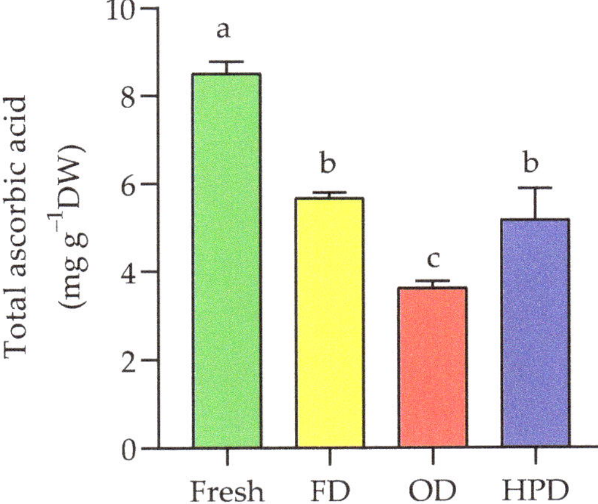

Figure 4. Total ascorbic acid content of *U. dioica* collected fresh or subjected to three different drying methods. FD: Freeze-drying; OD: Oven-drying; HPD: Heat pump drying. Means keyed with a different letter are significantly different for $p = 0.05$ following LSD test.

4. Conclusions

The presented data clearly indicate that the best dehydration technique is HPD since the resulting aqueous extract exhibited a high content in total phenols and a high antioxidant activity, as well as considerable levels in some phenolic compounds such as caffeoylmalic acid and caffeoylmalic acid dimer. Besides, the OD technique resulted in a high reduction in the ascorbic acid content (55%), and a lower decrease in this antioxidant compound was found when using the FD and HPD techniques compared with the reduction showed using the OD technique. Therefore, all three dehydration techniques appear to be efficient in favoring the extraction of phenols in water but adverse for ascorbic acid. Few studies have reported the quantification of total ascorbic acid in stinging nettle leaves in the literature; thus, this study can implement the knowledge of the quantification of this bioactive compound as well as show the loss of total ascorbic acid after the use of drying methods on plant leaves.

Author Contributions: Conceptualization, L.I. and A.P.; methodology, C.N. and L.M.G.; validation, C.C. and L.M.G.; formal analysis, C.N. and L.M.G.; investigation, C.N. and L.M.G.; resources, L.G. and A.P.; data curation, C.C.; writing—original draft preparation, L.M.G. and C.C.; writing—review and editing, L.D.B., A.P., L.I. and L.G.; supervision, L.G. and A.P.; project administration, L.G. All authors have read and agreed to the published version of the manuscript.

Funding: This research was performed in the framework of the "Erbi Boni" project, funded by FEASR Reg. UE 1305/2013—PSR 2014/2020—Regione Toscana Mis. 19—GAL MontagnAppennino Mis. 16.2.

Institutional Review Board Statement: Not applicable.

Informed Consent Statement: Not applicable.

Data Availability Statement: Data is contained within the article.

Acknowledgments: The authors are grateful to Diletta Piccotino for the useful help in the analysis and to Franca Bernardi, the owner of "Il Corniolo" farmhouse, for her availability and her valuable help during the sampling and the dehydration.

Conflicts of Interest: The authors declare no conflict of interest.

References

1. Dhouibi, R.; Affes, H.; Salem, M.B.; Hammami, S.; Sahnoun, Z.; Zeghal, K.M.; Ksouda, K. Screening of pharmacological uses of *Urtica dioica* and others benefits. *Prog. Biophys. Mol. Biol.* **2020**, *150*, 67–77. [CrossRef]
2. Kregiel, D.; Pawlikowska, E.; Antolak, H. *Urtica* spp.: Ordinary plants with extraordinary properties. *Molecules* **2018**, *23*, 1664. [CrossRef] [PubMed]
3. Grauso, L.; de Falco, B.; Lanzotti, V.; Motti, R. Stinging nettle, *Urtica dioica* L.: Botanical, phytochemical and pharmacological overview. *Phytochem. Rev.* **2020**, *19*, 1341–1377. [CrossRef]
4. Coklar, H.; Akbulut, M.; Kilinc, S.; Yildirim, A.; Alhassan, I. Effect of freeze, oven and microwave pretreated oven drying on color, browning index, phenolic compounds and antioxidant activity of hawthorn (*Crataegus orientalis*) fruit. *Not. Bot. Horti Agrobot. Cluj-Napoca* **2018**, *46*, 449–456. [CrossRef]
5. An, K.; Zhao, D.; Wang, Z.; Wu, J.; Xu, Y.; Xiao, G. Comparison of different drying methods on Chinese ginger (*Zingiber officinale* Roscoe): Changes in volatiles, chemical profile, antioxidant properties, and microstructure. *Food Chem.* **2016**, *197*, 1292–1300. [CrossRef] [PubMed]
6. Meng, Q.; Fan, H.; Li, Y.; Zhang, L. Effect of drying methods on physico-chemical properties and antioxidant activity of *Dendrobium officinale*. *J. Food Meas.* **2018**, *12*, 1–10. [CrossRef]
7. Perera, C.O.; Shafiur, R.M. Heat pump dehumidifier drying of food. *Trends Food Sci. Technol.* **1997**, *8*, 75–79. [CrossRef]
8. Ong, S.P.; Law, C.L. Drying Kinetics and Antioxidant Phytochemicals Retention of Salak Fruit under Different Drying and Pretreatment Conditions. *Dry. Technol.* **2011**, *29*, 429–441. [CrossRef]
9. Phoungchandang, S.; Wichayawadee, S.; Borwonsak, L. Kaffir lime leaf (*Citrus hystric* DC.) drying using tray and heat pump dehumidified drying. *Dry. Technol.* **2008**, *26*, 1602–1609. [CrossRef]
10. Hawlader, M.N.A.; Perera, C.O.; Tian, M.; Yeo, K.L. Drying of guava and papaya: Impact of different drying methods. *Dry. Technol.* **2006**, *24*, 77–87. [CrossRef]

11. Niruthchara, S.; Srzednicki, G.; Craske, J. Effects of drying treatments on the composition of volatile compounds in dried nectarines. *Dry. Technol.* **2007**, *25*, 877–881.
12. Dewanto, V.; Adom, K.K.; Liu, R.H. Thermal processing enhances the nutritional value of tomatoes by increasing total antioxidant activity. *J. Agric. Food Chem.* **2002**, *50*, 3010–3014. [CrossRef] [PubMed]
13. Brand Williams, W.; Cuvelier, M.E.; Berset, C. Use of a free radical method to evaluate antioxidant activity. *Lebensm. Wiss. Technol.* **1995**, *28*, 25–30. [CrossRef]
14. Benzie, I.F.; Strain, J.J. The ferric reducing ability of plasma (FRAP) as a measure of "antioxidant power": The FRAP assay. *Anal. Biochem.* **1996**, *239*, 70–76. [CrossRef] [PubMed]
15. Zhou, X.; Chan, S.W.; Tseng, H.L.; Deng, Y.; Hoi, P.M.; Choi, P.S.; Or, P.M.; Yang, J.M.; Lam, F.F.; Lee, S.M.; et al. Danshensu is the major marker for the antioxidant and vasorelaxation effects of Danshen (*Salvia miltiorrhiza*) water-extracts produced by different heat water-extractions. *Phytomedicine* **2012**, *19*, 1263–1269. [CrossRef]
16. Negro, C.; Sabella, E.; Nicolì, F.; Pierro, R.; Materazzi, A.; Panattoni, A.; Aprile, A.; Nutricati, E.; Vergine, M.; Miceli, A.; et al. Biochemical Changes in Leaves of *Vitis vinifera* cv. Sangiovese Infected by Bois Noir Phytoplasma. *Pathogens* **2020**, *9*, 269. [CrossRef]
17. Kampfenkel, K.; Van Montagu, M.; Inzé, D. Extraction and determination of ascorbate and dehydroascorbate from plant tissue. *Anal. Biochem.* **1995**, *225*, 165–167. [CrossRef]
18. Blank, D.E.; Bellaver, M.; Fraga, S.; Lopes, T.J.; de Moura, N.F. Drying kinetics and bioactive compounds of *Bunchosia glandulifera*. *J. Food Process Eng.* **2018**, *41*, e12676. [CrossRef]
19. López-Vidaña, E.C.; Pilatowsky Figueroa, I.; Cortés, F.B.; Rojano, B.A.; Ocaña, A.N. Effect of temperature on antioxidant capacity during drying process of mortiño (*Vaccinium meridionale* Swartz). *Int. J. Food Prop.* **2017**, *20*, 294–305. [CrossRef]
20. Farag, M.A.; Weigend, M.; Luebert, F.; Brokamp, G.; Wessjohann, L.A. Phytochemical, phylogenetic, and anti-inflammatory evaluation of 43 *Urtica* accessions (stinging nettle) based on UPLC–Q-TOF-MS metabolomic profiles. *Phytochemistry* **2013**, *96*, 170–183. [CrossRef]
21. Komes, D.; Belščak-Cvitanović, A.; Horžić, D.; Rusak, G.; Likić, S.; Berendika, M. Phenolic composition and antioxidant properties of some traditionally used medicinal plants affected by the extraction time and hydrolysis. *Phytochem. Anal.* **2011**, *22*, 172–180. [CrossRef] [PubMed]
22. Pinelli, P.; Ieri, F.; Vignolini, P.; Bacci, L.; Baronti, S.; Romani, A. Extraction and HPLC analysis of phenolic compounds in leaves, stalks, and textile fibers of *Urtica dioica* L. *J. Agric. Food Chem.* **2008**, *56*, 9127–9132. [CrossRef] [PubMed]
23. Giada, M.D.L.R. Food phenolic compounds: Main classes, sources and their antioxidant power. Oxidative stress and chronic degenerative diseases—A role for antioxidants. *InTech* **2013**, 87–112. [CrossRef]
24. Singleton, V.L.; Orthofer, R.; Lamuela-Raventós, R.M. Analysis of total phenols and other oxidation substrates and antioxidants by means of Folin-Ciocalteu reagent. *Meth. Enzymol.* **1999**, *299*, 152–178.
25. Barros, L.; Dueñas, M.; Carvalho, A.M.; Ferreira, I.C.; Santos-Buelga, C. Characterization of phenolic compounds in flowers of wild medicinal plants from Northeastern Portugal. *Food Chem. Toxicol.* **2012**, *50*, 1576–1582. [CrossRef]
26. Martins, N.; Barros, L.; Santos-Buelga, C.; Ferreira, I.C. Antioxidant potential of two Apiaceae plant extracts: A comparative study focused on the phenolic composition. *Ind. Crops Prod.* **2016**, *79*, 188–194. [CrossRef]
27. Zeković, Z.; Cvetanović, A.; Švarc-Gajić, J.; Gorjanović, S.; Sužnjević, D.; Mašković, P.; Savic, S.; Radojkovic, M.; Đurović, S. Chemical and biological screening of stinging nettle leaves extracts obtained by modern extraction techniques. *Ind. Crops Prod.* **2017**, *108*, 423–430. [CrossRef]
28. Shonte, T.T.; Duodu, K.G.; de Kock, H.L. Effect of drying methods on chemical composition and antioxidant activity of underutilized stinging nettle leaves. *Heliyon* **2020**, *6*, e03938. [CrossRef]
29. Carvalho, A.R.; Costa, G.; Figueirinha, A.; Liberal, J.; Prior, J.A.; Lopes, M.C.; Cruz, M.T.; Batista, M.T. *Urtica* spp.: Phenolic composition, safety, antioxidant and anti-inflammatory activities. *Food Res. Intern.* **2017**, *99*, 485–494. [CrossRef]
30. Hudec, J.; Burdová, M.; Kobida, L.U.; Komora, L.; Macho, V.; Kogan, G.; Turianica, I.; Kochanová, R.; Ložek, O.; Habán, M.; et al. Antioxidant capacity changes and phenolic profile of *Echinacea purpurea*, nettle (*Urtica dioica* L.), and dandelion (*Taraxacum officinale*) after application of polyamine and phenolic biosynthesis regulators. *J. Agric. Food Chem.* **2007**, *55*, 5689–5696. [CrossRef]
31. Bai, Y.P.; Zhou, H.M.; Zhu, K.R.; Li, Q. Effect of thermal treatment on the physicochemical, ultrastructural and nutritional characteristics of whole grain highland barley. *Food Chem.* **2020**, 128657. [CrossRef]
32. Lim, Y.Y.; Murtijaya, J. Antioxidant properties of *Phyllanthus amarus* extracts as affected by different drying methods. *LWT-Food Sci. Technol.* **2007**, *40*, 1664–1669. [CrossRef]
33. Grevsen, K.; Frette, X.; Christensen, L.P. Concentration and composition of flavonol glycosides and phenolic acids in areal parts of stinging nettle (*Urtica dioica* L.) are affect by high nitrogen fertilization and harvest time. *Eur. J. Hortic. Sci.* **2008**, *73*, 20–27.
34. Orčić, D.; Francišković, M.; Bekvalac, K.; Svirčev, E.; Beara, I.; Lesjak, M.; Mimica-Dukić, N. Quantitative determination of plant phenolics in *Urtica dioica* extracts by high-performance liquid chromatography coupled with tandem mass spectrometric detection. *Food Chem.* **2014**, *143*, 48–53. [CrossRef]
35. Vajić, U.; Grujić-Milanović, J.; Živković, J.; Šavikin, K.; Gođevac, D.; Miloradović, Z.; Bugarski, B.; Mihailović-Stanojević, N. Optimization of extraction of stinging nettle leaf phenolic compounds using response surface methodology. *Ind. Crops Prod.* **2015**, *74*, 912–917.

36. Klava, D.; Kampuse, D.; Tomsone, L.; Kince, T.; Ozola, L. Effect of drying technologies on bioactive compounds maintenance in pumpkin by-products. *Agron. Res.* **2018**, *16*, 1728–1741.
37. Parsons, H.T.; Fry, S.C. Oxidation of dehydroascorbic acid and 2, 3-diketogulonate under plant apoplastic conditions. *Phytochemistry* **2012**, *75*, 41–49. [CrossRef]

Article

Indices for the Assessment of Glycoalkaloids in Potato Tubers Based on Surface Color and Chlorophyll Content

Shimeles Tilahun [1,2,3], Hee Sung An [1], Tifsehit Solomon [4], Min Woo Baek [1], Han Ryul Choi [1], Hee Cheol Lee [1] and Cheon Soon Jeong [1,*]

1 Department of Horticulture, Kangwon National University, Chuncheon, Gangwon 24341, Korea; shimeles@kangwon.ac.kr (S.T.); ahs@ht-calbee.co.kr (H.S.A.); minwoo100@kangwon.ac.kr (M.W.B.); hanryul192@kangwon.ac.kr (H.R.C.); yaya9459@kangwon.ac.kr (H.C.L.)
2 Agriculture and Life Science Research Institute, Kangwon National University, Chuncheon 24341, Korea
3 Department of Horticulture and Plant Sciences, Jimma University, Jimma 378, Ethiopia
4 Department of Biology, Wollega University, Nekemte 395, Ethiopia; tifsehits@gmail.com
* Correspondence: jeongcs@kangwon.ac.kr; Tel.: +82-033-2506409

Received: 26 November 2020; Accepted: 14 December 2020; Published: 16 December 2020

Abstract: Glycoalkaloids (GAs) are toxic to humans at higher concentrations. However, studies also suggest the health benefits of GAs depending on the dose and conditions of use. Methods that have been used to determine GA content in potato tubers are destructive and time-consuming and require skilled personnel and high-performance laboratory equipment. We conducted this study to develop indices for the prediction of the level of total GAs in potato tubers at different greening stages based on surface color readings and chlorophyll (Chl) development. Color values (Hunter L*, a*, b*, a*/b*), Chls (Chl a, Chl b, and total Chls) and GA (α-solanine, α-chaconine, and total GAs) content were measured from tubers of 'Atlantic' and 'Trent' potato cultivars at three-week intervals in up to six greening stages during the storage at room conditions (22 °C, 12-h shift of light-dark cycles). The results have revealed that greening, Chls, and GA content significantly increased for the two cultivars as the stage proceeded. The toxic level of GAs (>200 mg kg^{-1} FW) was accumulated at the late greening stages, accompanied by the highest Chl content. Finally, indices were developed based on surface color and Chl content for estimation of the safe GA levels for the consumption of the two commercially and commonly used potato cultivars. Moreover, the developed indices could be used as basic information to adapt to other potato cultivars.

Keywords: α-solanine; α-chaconine; color variables; chlorophyll contents; color index

---

1. Introduction

Potato (*Solanum tuberosum* L.) is the leading non-grain food commodity in the global food system, with production reaching 368.2 million Mt from 17.6 million ha in 2018 [1]. It is the fourth most important food crop in the world, following maize, wheat, and rice. The Republic of Korea produced 553,596 Mt of potato from 25,772 ha of land in 2018 [1]. Potatoes play a fundamental part in the effort made to ensure food and nutritional security for the increasing world population [2,3]. They are one of the most efficient crops for converting natural resources, labor, and capital into high-quality food. The production of potato is relatively easy, and their genetic complexity allows a diversity of genotypes for any climate, culture, and conditions [3]. They produce more nutritious food in a short growing cycle on small land and in harsher climates than any other major crop [2].

Potato represents half of the root and tuber crops consumed for a carbohydrate food source in the world [4]. Potatoes are rich in carbohydrates, which makes them a good source of energy [2]. They have the highest protein content, around 2.1 percent on a fresh weight basis, in the family of root and tuber crops and protein of high quality, with an amino-acid pattern that matches human requirements [2,3]. Additionally, potatoes are sources of a variety of essential vitamins and minerals, such as vitamins C and B6 and the minerals potassium, magnesium, and iron [2,3]. A single medium-sized potato contains about half the recommended daily intake of vitamin C and a fifth of the recommended daily value of potassium [2,3]. Potatoes have been traditionally consumed after boiling, frying, or baking as a main dish or with other food types. Different popular snack products, such as crisps or French fries, that appeal to consumers, due to appearance, texture, or flavor, are also processed from potato tubers [3,5].

The potato belongs to the *Solanaceae* family, which is known for producing secondary natural poisonous metabolites called glycoalkaloids (GAs) [5]. GAs play an important defensive role due to their toxic nature against pests [5]. α-solanine and α-chaconine are the two major glycoalkaloids in cultivated potatoes that together account for 95% of the total glycoalkaloid content [6,7]. In the tubers, there is a higher concentration of GAs obtained in the skin, around the eyes, wounded areas, and in the sprouts [8,9]. Exposing potato tubers to light in storage or at home causes greening, and greened tubers accumulate GAs [10]. The processes of greening and GA accumulation are concurrent but independent; the greening shows the formation of chlorophylls (Chls) and it is also considered as an indication for an increase in the level of GAs [11]. GA synthesis can also be elevated by damage to the tubers during post-harvest operations [12].

GAs are toxic to humans at higher concentrations. The safe upper limit for a human is 200 mg kg^{-1} total GAs of fresh tuber weight [13]. Potato tubers that contain over 200 mg kg^{-1} total GAs of fresh tuber weight possess a bitter off-flavor and may cause gastro-enteric symptoms, comas, and even death [14–18]. Symptoms of toxicity include headache, nausea, fatigue, vomiting, abdominal pain, diarrhea, apathy, restlessness, drowsiness, mental confusion, rambling, incoherence, stupor, hallucinations, trembling, and visual disturbances [5,12,14,19]. However, studies suggest that GAs have also health benefits, such as anticancer, antimalarial, anti-inflammatory, hypoglycemic, and hypo-cholesterol emic activities, depending on the dose and conditions of use [14,20]. Therefore, it is crucial to develop an indicative index that can be used easily to avoid the potential toxicity of GAs to human beings.

The different methods that have been used to determine the GAs are time-consuming and destructive and require skilled personnel and high-performance laboratory equipment. It is difficult for producers and consumers to determine the safe level of alkaloids with the existing methods of determination. Therefore, the current study was designed to develop indices for the prediction of the level of GAs in potato tubers at different greening stages based on surface color readings and Chl development.

2. Materials and Methods

2.1. Plant Material

Commercially and commonly used 'Atlantic' and 'Trent' potato cultivars were selected for this study. Tubers of the two cultivars were obtained from the Haitai-Calbee snack factory, Korea. Relatively uniform-sized defect-free tubers were selected for the experiment and stored at room conditions (22 °C, 12-h shift of light-dark cycles) to simulate the consumers' practice and to allow greening. During sampling, 20 tubers ('Atlantic' and 'Trent'; 10 tubers each) were used for each greening stage. Subsampling was done at six different greening stages at three-week intervals for up to 15 weeks of storage, and representative photos of the tubers from both cultivars were acquired at each greening stage. Samples for the analysis of GAs and Chl content were prepared immediately after taking representative photos and the color reading from the surface of intact tubers. The samples were then frozen by liquid nitrogen and stored in a deep freezer (−80 °C) until analysis [21].

2.2. Color Measurement and Analysis

Intact and undamaged tubers were used to take surface readings of Hunter L*, a*, and b* color variables using CR-400 chroma meter (Minolta, Tokyo, Japan). The Hunter a* represents a chromatic redness parameter that ranges from red to green with positive to negative values, respectively. The Hunter b* represents a chromatic yellowness parameter that ranges from yellow with positive values to blue with negative values, while the Hunter L* represents the degree of brightness, which ranges from black with 0 value to white with 100 value [22]. The color variables were measured and recorded from the surface of each tuber five times and the values were averaged. The color was measured at three-week intervals until the 15th week of storage. During sampling, 20 tubers ('Atlantic' and 'Trent'; 10 tubers each) were used for each greening stage. Hence, a total of 120 tubers (60 tubers from each) were used for the study.

2.3. Extraction and Quantification of Chls and GAs

Chls and GAs were extracted from 20 potato tubers at each stage (10 tubers from each cultivar) for a 15-weeks storage period at three-week intervals. The peripheral 10 mm tissue was sampled, since Chl and GAs are accumulated in the outer cell layer of the tuber [8,9,23]. Dimethyl sulfoxide (DMSO) chlorophyll extraction procedure was used, as described by [24]. The readings were measured at 645 nm and 663 nm using a spectrophotometer (Thermo Fisher Scientific, Waltham, MA, USA) against a DMSO blank. Subsequently, Chl a, Chl b, and total Chls were calculated by Arnon [25] using equations as follows:

a. Chl a (mg g^{-1}) peel fresh weight) = [(12.7*A663) − (2.69*A645)]*(V/1000*W)
b. Chl b (mg g^{-1}) peel fresh weight) = [(22.9*A645) − (4.68*A663)]*(V/1000*W)
c. Total Chls = Chl a + Chl b

V = volume of solvent; W = fresh weight of the extracted tissue.

Extraction of GAs was made as stated by Tilahun et al. [9] and quantification of α-solanine and α-chaconine was done by ultra-performance liquid chromatography-tandem mass spectrometry (UPLC-MS/MS) [26]. The spectrometer was adjusted, as described by Tilahun et al. [9], Zywicki et al. [26], and Nie et al. [27], for detection of α-solanine and α-chaconine.

2.4. Procedure to Develop Indices for the Estimation of Total GAs

The photos of representative potato tubers from both 'Atlantic' and 'Trent' potato cultivars were acquired by a Canon digital camera (EOS 200D, Tokyo, Japan) at the six different greening stages with three-week intervals, after peeling a portion of peripheral 10 mm tissue to show the extent of greening. The average data obtained from the color measurements, total Chls, and total GAs were then matched to the corresponding photo at each stage based on correlation coefficients of color values and Chls content vs. total GAs.

2.5. Statistical Analysis

The experiment was conducted in a completely randomized design. The data were subjected to analysis of variance (ANOVA) to determine the significance of differences between cultivars and greening stages at $p < 0.05$ using SAS statistical software (SAS/STAT® 9.1; SAS Institute Inc., Cary, NC, USA), and the Pearson correlation test was used to correlate the collected parameters.

3. Results and Discussion

3.1. Color Variables and GAs

The current study results have revealed a significant difference for Hunter L*, a*, b*, and a*/b* values among the six stages of greening for both 'Atlantic' and 'Trent' potato cultivars. The difference between the two cultivars was also significant, though not at all greening stages (Table 1). The Hunter L* values decreased across greening stages 1–6, indicating the decrease in tubers' brightness for both cultivars. Grunenfelder et al. [28] also reported a decrease in L* values of tubers as the storage days proceeded from 0–7 days. Hunter a* values, which indicate the chromatic value ranging from red (+a*) to green (−a*), have also shown a decreasing trend from stage 1–6, indicating an increase in the greening of the tubers for both cultivars [22]. The hunter a* values were positive for tubers in stages 1 and 2, and negative hunter a* values were recorded for tubers in the remaining stages. A study conducted by Haase [29] has also shown a significant change of tuber color towards green because of exposure of potato tubers to light. Similar to Hunter a* values, a decreasing trend and significant differences were observed in the Hunter b* values among the six stages for both cultivars. The Hunter a*/b* ratios were significantly different and follow the same trend as Hunter a* color values among the greening stages of both cultivars. Tilahun et al. [30] also reported the same trends of Hunter a* values and a*/b* ratios during the evaluation of color changes in tomatoes.

Table 1. Hunters L*, a*, b*, and a*/b* color values of 'Atlantic' and 'Trent' potato cultivars under six greening stages during storage at room conditions (22 °C) with a 12-h shift of light-dark cycles.

Parameters	Cultivars	Greening Stages					
		1	2	3	4	5	6
L*	Atlantic	58.30_{a}^{A}	56.74_{bc}^{B}	55.66_{bcd}^{A}	55.37_{bcd}^{A}	55.29_{cd}^{A}	54.30_{d}^{A}
	Trent	56.88_{b}^{B}	58.93_{a}^{A}	55.95_{bc}^{A}	55.84_{bc}^{A}	56.28_{bc}^{A}	52.64_{d}^{B}
a*	Atlantic	1.83_{a}^{A}	0.99_{bc}^{A}	-1.23_{cd}^{A}	-1.90_{de}^{A}	-2.20_{de}^{A}	-2.89_{f}^{A}
	Trent	2.10_{a}^{A}	1.01_{bc}^{A}	0.08_{cd}^{B}	-0.55_{de}^{B}	-0.28_{de}^{B}	-3.78_{f}^{A}
b*	Atlantic	26.10_{ab}^{B}	25.21_{bc}^{B}	24.86_{c}^{B}	25.82_{abc}^{B}	23.72_{d}^{B}	22.46_{e}^{A}
	Trent	29.22_{a}^{A}	28.32_{ab}^{A}	27.16_{c}^{A}	27.81_{bc}^{A}	26.41_{cd}^{A}	22.76_{e}^{A}
a*/b*	Atlantic	0.070_{a}^{A}	0.039_{bc}^{A}	-0.049_{cd}^{B}	-0.076_{de}^{B}	-0.093_{de}^{B}	-0.129_{f}^{A}
	Trent	0.072_{a}^{A}	0.036_{bc}^{A}	0.003_{cd}^{A}	-0.020_{de}^{A}	-0.011_{de}^{A}	-0.166_{f}^{B}

The mean with different uppercase letters within the same column is significantly different ($p < 0.05$), while the mean with different lowercase letters within the same row is significantly different ($p < 0.05$).

A significant difference was also observed in GA content among the six stages for the two cultivars (Figure 1). The α-solanine, α-chaconine, and total GA contents increased as it went from stage 1–6. The α-solanine and α-chaconine content were cultivar-dependent in the present study. α-solanine was lower than α-chaconine in the 'Atlantic' cultivar throughout the greening stages. However, the ratio of α-solanine to α-chaconine was nearly 1:1 in the 'Trent' cultivar. Tajner-Czopek [31] reported cultivar-dependent ratios of α-solanine to α-chaconine, ranging from 1:1.9 to 1:2.5 for early potato cultivars. In this study, both cultivars accumulated more than the recommended safe level (>200 mg/kg) of glycoalkaloids at the late stages [7,9,13]. 'Atlantic' accumulated a toxic level of total GAs (296 mg kg^{-1}) at the sixth stage with a Hunter a* value of −2.89, indicating the tuber color change into green. 'Trent' accumulated a toxic level (205.38 mg kg^{-1}) of GAs at stage 5, earlier than 'Atlantic', and reached (225.96 mg kg^{-1}) stage 6 with Hunter a* values of −0.28 and −3.78, respectively (Table 1 and Figure 2). Hunter L* values showed a decrease in the brightness of the tuber at the late stages (Table 1).

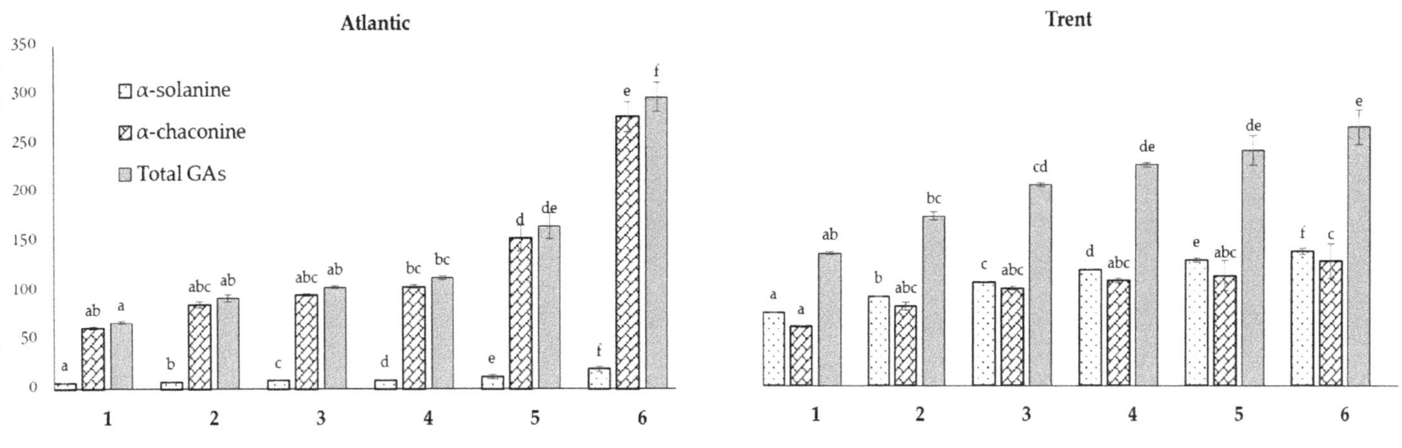

Figure 1. Glycoalkaloids content of 'Atlantic' and 'Trent' potato cultivars under six greening stages during storage at room conditions (22 °C) with a 12-h shift of light-dark cycles. The bars with different letters indicate a significant difference ($p < 0.05$) between the greening stages. The vertical bars represent the standard error of the mean (n = 10).

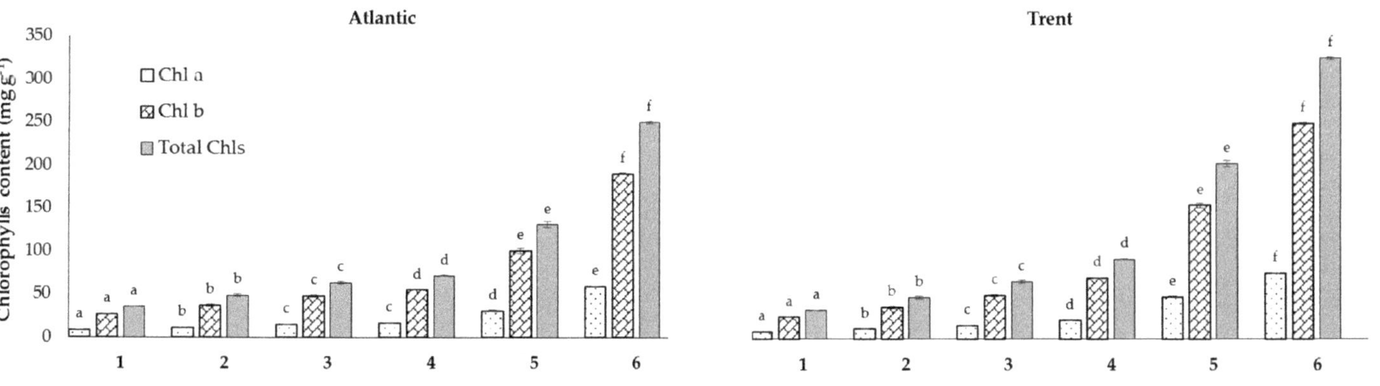

Figure 2. Chlorophyll content of 'Atlantic' and 'Trent' potato cultivars under six greening stages during storage at room conditions (22 °C) with a 12-h shift of light-dark cycles. The bars with different letters indicate a significant difference ($p < 0.05$) between the greening stages. The vertical bars represent the standard error of the mean (n = 10).

3.2. Chls and GAs

Exposure of potato tubers to light causes greening due to the formation of Chls and accumulation of toxic GAs [10,11]. The results of this study have revealed that Chl and GA content were significantly different among the six greening stages for both 'Atlantic' and 'Trent' potato cultivars. An increasing trend in total Chl and total GA content was observed as the greening stages went from stage 1 to 6 for both cultivars (Figures 1 and 2). Similar results were reported by Percival [32], in which Chl and total GA concentrations steadily increased in three potato cultivars in response to light over time. Okamoto et al. [33] have also shown the dependence of tuber Chl and total GA accumulation on light and confirmed the major role of light in both greening and GAs synthesis. They also observed no Chls under the absence of light, while there was an increase in accumulation of Chls under high light intensity. In the present study, the highest Chl a, Chl b, and total Chl content were recorded on the sixth stage for both 'Atlantic' and 'Trent' cultivars (Figure 1).

The total Chls increased from 35.71 and 33.57 mg g^{-1} in the first stage to 249.86 and 326.86 mg g^{-1} in the sixth stage for 'Atlantic' and 'Trent' cultivars, respectively. Concurrently, the total glycoalkaloids increased from 67.52 and 115.63 mg kg^{-1} in the first stage to 296.9 and 225.96 mg kg^{-1} in the sixth stage for 'Atlantic' and 'Trent' cultivars, respectively. The safe level of total glycoalkaloid content accepted for consumption (200 mg kg^{-1}) was recorded at stage 1–5 for 'Atlantic' and stage 1–4 for 'Trent'. 'Atlantic' accumulated the toxic level of glycoalkaloids at stage 6 and 'Trent' at stages 5 and 6. Therefore, it is safe to consume the potato tubers of both cultivars up to greening stage 4 (Figures 3 and 4).

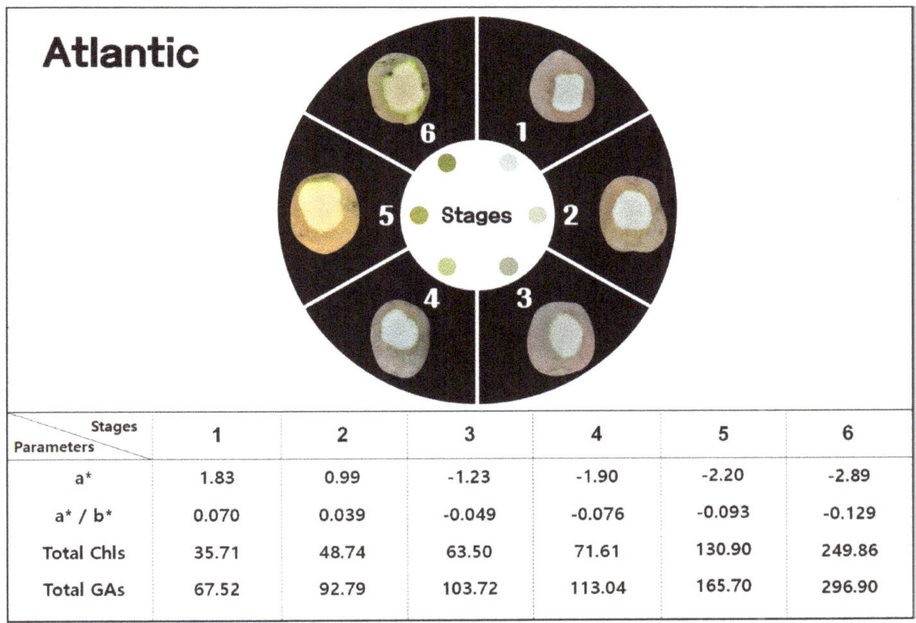

Parameters \ Stages	1	2	3	4	5	6
a*	1.83	0.99	-1.23	-1.90	-2.20	-2.89
a* / b*	0.070	0.039	-0.049	-0.076	-0.093	-0.129
Total Chls	35.71	48.74	63.50	71.61	130.90	249.86
Total GAs	67.52	92.79	103.72	113.04	165.70	296.90

Figure 3. Index for the estimation of total glycoalkaloids (Gas) of 'Atlantic' potato cultivar at six greening stages based on color values and chlorophyll (Chl) content.

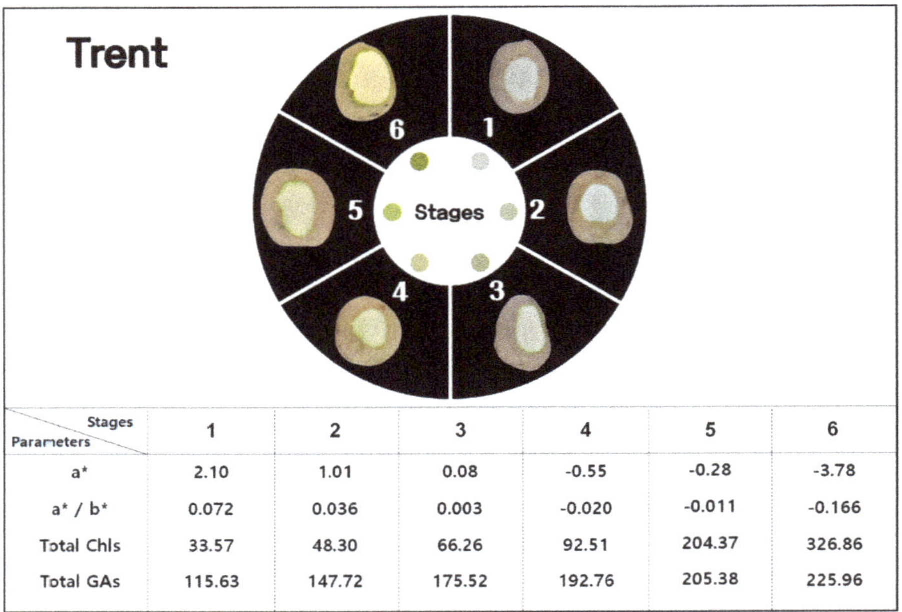

Parameters \ Stages	1	2	3	4	5	6
a*	2.10	1.01	0.08	-0.55	-0.28	-3.78
a* / b*	0.072	0.036	0.003	-0.020	-0.011	-0.166
Total Chls	33.57	48.30	66.26	92.51	204.37	326.86
Total GAs	115.63	147.72	175.52	192.76	205.38	225.96

Figure 4. Index for the estimation of total GAs of 'Trent' potato cultivar at six greening stages based on color values and Chl content.

3.3. Correlation of the Evaluated Parameters and Indices for the Estimation of Total GAs

The correlation result between all the color values vs. the total GAs showed a significant inverse relationship, except for the relationship between Hunter b* and total GAs (Table 2). The Hunter L* color values and total GA correlation (−0.62) indicated a decrease in brightness as the total glycoalkaloid content increased. Correspondingly, the significant correlation (−0.65) between the Hunter a* value and total GA content showed an increase in the greening of the tubers as the glycoalkaloids accumulation increased. The significant correlation (−0.67) between the Hunter a*/b* ratio and total GA content implies that a*/b* ratio could be a good indicator of GA content. In agreement with our results, the previous studies by Arias et al. [34], Helyes et al. [35], and Tilahun et al. [30] reported the Hunter a* and a*/b* ratio as good indicators of lycopene content (redness) in tomatoes. Conversely, the correlation between Chl a, Chl b, and total Chl vs. total GA in the current study showed a highly significant positive relationship, indicating an increase in total GA content as the total Chl content increased (Table 2). Total Chl content showed the highest (r = 0.84) correlation to the total GA contents. This finding agrees with Spoladore et al. [36], who reported a strong positive relationship between Chl and GA content in potato tubers. In summary, the overall results of the evaluated parameters indicated that the Hunter a* values, a*/b* ratio, and Chl content could be used as the key indicators of the total GA contents. Hence, we used the above parameters to develop indices for the estimation of total GA content, as shown in Figures 3 and 4.

Table 2. Correlation between color values, Chls, and GA contents of 'Atlantic' and 'Trent' potato cultivars under six greening stages during storage at room conditions (22 °C) with a 12-h shift of light-dark cycles.

	Hunter L*	Hunter a*	Hunter b*	a*/b*	Chl a	Chl b	Total Chl	α-solanine	α-chaconine	Total GA
Hunter L*	1	0.89 ***	0.73 **	0.90 ***	−0.78 **	−0.79 **	−0.80 **	−0.16 ns	−0.55 ns	−0.62 *
Hunter a*		1	0.82 ***	0.99 ***	−0.78 **	−0.78 **	−0.82 ***	−0.05 ns	−0.67 *	−0.65 **
Hunter b*			1	0.83 ***	−0.67 *	−0.67 *	−0.69 *	0.35 ns	−0.71 **	−0.40 ns
a*/b*				1	−0.80 **	−0.80 **	−0.84 ***	−0.06 ns	−0.69 *	−0.67 **
Chl a					1	0.99 ***	0.99 ***	0.39 ns	0.60 *	0.82 **
Chl b						1	0.99 ***	0.39 ns	0.59 *	0.81 **
Total Chl							1	0.38 ns	0.63 *	0.84 ***
α−solanine								1	−0.25 ns	0.47 ns
α−chaconine									1	0.74**
Total GA										1

ns, *, **, and *** indicate non-significant and significant differences at $p < 0.05$, 0.01, and 0.001, respectively.

4. Conclusions

The current study has tried to develop simple indices to detect the toxic level of GAs in potato tubers using surface color and Chl development. Potato tubers from 'Atlantic' and 'Trent' cultivars were used to determine the greening stages at which toxic levels of GAs can be accumulated. The greening, Chls, and GA contents significantly increased for the two cultivars as the stage proceeded. The toxic level of GAs was accumulated at the late greening stages of the tubers, accompanied by the highest Chl content. Taken together, indices were developed based on surface color and Chl content for estimation of the safe GA levels for the consumption of the two commercially and commonly used potato cultivars. The developed indices can be used easily to avoid the potential toxicity of GAs to human beings. Moreover, the developed indices could be used as basic information to adapt it to other potato cultivars.

Author Contributions: Conceptualization, S.T., H.S.A., and C.S.J.; methodology, S.T. and H.S.A.; execution of experiment, S.T., M.W.B., and H.C.L.; software, S.T. and T.S.; formal analysis, S.T., T.S., and H.R.C.; resources, C.S.J.; original draft preparation, T.S., H.S.A., and H.R.C.; review and editing S.T. and T.S.; supervision, C.S.J.; project administration, M.W.B.; funding acquisition, C.S.J. All authors have read and agreed to the published version of the manuscript.

Funding: Financial support from scientific and technological support program through the National Research Foundation of Korea (NRF) funded by the Ministry of Education (NRF-2019K1A3A9A0100002412) is gratefully acknowledged.

Conflicts of Interest: The authors declare no conflict of interest.

References

1. FAOSTAT Food & Agriculture Organization of the United Nations Statistics Division. Available online: http://faostat3.fao.org/home/index.html (accessed on 20 October 2020).
2. Prokop, S.; Albert, J. *Potatoes, Nutrition and Diet-International Year of the Potato*; Food and Agriculture Organization of the United Nations Factsheet: Roma, Italy, 2008; pp. 1–2.
3. Hussain, T. Potatoes: Ensuring Food for the Future. *Adv. Plants Agric. Res.* **2016**, *3*, 178–182. [CrossRef]
4. Beals, K.A. Potatoes, Nutrition and Health. *Am. J. Potato Res.* **2019**, *96*, 102–110. [CrossRef]
5. Omayio, D.G.; Abong, G.O.; Okoth, M.W. A review of occurrence of glycoalkaloids in potato and potato products. *Curr. Res. Nutr. Food Sci.* **2016**, *4*, 195–202. [CrossRef]
6. Shen, K.H.; Liao, A.C.H.; Hung, J.H.; Lee, W.J.; Hu, K.C.; Lin, P.T.; Liao, R.F.; Chen, P.S. α-Solanine inhibits invasion of human prostate cancer cell by suppressing epithelial-mesenchymal transition and MMPs expression. *Molecules* **2014**, *19*, 11896–11914. [CrossRef] [PubMed]
7. Tilahun, S.; An, H.S.; Hwang, I.G.; Choi, J.H.; Baek, M.W.; Choi, H.R.; Park, D.S.; Jeong, C.S. Prediction of α-Solanine and α-Chaconine in Potato Tubers from Hunter Color Values and VIS/NIR Spectra. *J. Food Qual.* **2020**. [CrossRef]
8. Kirui, G.K.; Misra, A.K.; Olanya, O.M.; Friedman, M.; El-Bedewy, R.; Ewell, P.T. Glycoalkaloid content of some superior potato (*Solanum tuberosum* L.) clones and commercial cultivars. *Arch. Phytopathol. Plant Prot.* **2009**, *42*, 453–463. [CrossRef]

9. Tilahun, S.; An, H.S.; Choi, H.R.; Park, D.S.; Lee, J.S.; Jeong, C.S. Effects of Storage Temperature on Glycoalkaloid Content, Acrylamide Formation, and Processing-related Variables of Potato Cultivars. *Agric. Life Environ. Sci.* **2019**, *31*, 128–142.
10. Jakuczun, H.; Zimnoch-guzowska, E. Inheritance of tuber greening under light exposure in diploid potatoes. *Am. J. Potato Res.* **2006**, *83*, 211–221. [CrossRef]
11. Grunenfelder, L.A.; Knowles, L.O.; Hiller, L.K.; Knowles, N.R. Glycoalkaloid development during greening of fresh market potatoes (*Solanum tuberosum* L.). *J. Agric. Food Chem.* **2006**, *54*, 5847–5854. [CrossRef]
12. Prabhat, K.N.; Ramayya, N.; Duncan, E.; Niranjan, K. Potato glycoalkaloids: Formation and strategies for mitigation. *J. Sci. Food Agric.* **2008**, *88*, 1869–1881. [CrossRef]
13. Şengül, M.; Keleş, F.; Keleş, M.S. The effect of storage conditions (temperature, light, time) and variety on the glycoalkaloid content of potato tubers and sprouts. *Food Control.* **2004**, *15*, 281–286. [CrossRef]
14. Friedman, M. Potato glycoalkaloids and metabolites: Roles in the plant and in the diet. *J. Agric. Food Chem.* **2006**, *54*, 8655–8681. [CrossRef] [PubMed]
15. Cantwell, M. A Review of Important Facts about Potato Glycoalkaloids. *Perishables Handl. Newsl.* **1996**, *2*, 27.
16. Kodamatani, H.; Saito, K.; Niina, N.; Yamazaki, S.; Tanaka, Y. Simple and sensitive method for determination of glycoalkaloids in potato tubers by high-performance liquid chromatography with chemiluminescence detection. *J. Chromatogr. A* **2005**, *1100*, 26–31. [CrossRef]
17. Sotelo, A.; Serrano, B. High-performance liquid chromatographic determination of the glycoalkaloids α-solanine and α-chaconine in 12 commercial varieties of Mexican potato. *J. Agric. Food Chem.* **2000**, *48*, 2472–2475. [CrossRef] [PubMed]
18. Coxon, D.T.; Price, K.R.; Jones, P.G. A simplified method for the determination of total glycoalkaloids in potato tubers. *J. Sci. Food Agric.* **1979**, *30*, 1043–1049. [CrossRef]
19. Abu Bakar Siddique, M.; Brunton, N. Food Glycoalkaloids: Distribution, Structure, Cytotoxicity, Extraction, and Biological Activity. *Alkaloids Their Importance Nat. Hum. Life* **2019**. [CrossRef]
20. Uluwaduge, D.I.; Lecturer, S.; Lanka, S. Glycoalkaloids, bitter tasting toxicants in potatoes: A review. *Int. J. Food Sci. Nutr.* **2018**, *3*, 188–193.
21. Tilahun, S.; Seo, M.H.; Park, D.S.; Jeong, C.S. Effect of cultivar and growing medium on the fruit quality attributes and antioxidant properties of tomato (*Solanum lycopersicum* L.). *Hortic. Environ. Biotechnol.* **2018**, *59*. [CrossRef]
22. McGuire, R.G. Reporting of Objective Color Measurements. *HortScience* **1992**, *27*, 1254–1255. [CrossRef]
23. Jadhav, S.J.; Salunkhe, D.K. Formation and Control of Chlorophyll and Glycoalkaloids in Tubers of *Solanum tuberosum* L. and Evaluation of Glycoalkaloid Toxicity*. *Adv. Food Res.* **1975**, *21*, 307–354. [CrossRef]
24. Hiscox, J.D.; Israelstam, G.F. A method for the extraction of chlorophyll from leaf tissue without maceration. In Proceedings of the 2010 International Conference on E-Business and E-Government, Guangzhou, China, 7–9 May 2010; pp. 5474–5478. [CrossRef]
25. Arnon, D.I. Copper enzymes in isolated chloroplasts. Polyphenoloxidase in Beta vulgaris. *Plant Physiol.* **1949**, *24*, 1–15. [CrossRef]
26. Zywicki, B.; Catchpole, G.; Draper, J.; Fiehn, O. Comparison of rapid liquid chromatography-electrospray ionization-tandem mass spectrometry methods for determination of glycoalkaloids in transgenic field-grown potatoes. *Anal. Biochem.* **2005**, *336*, 178–186. [CrossRef]
27. Nie, X.; Zhang, G.; Lv, S.; Guo, H. Steroidal glycoalkaloids in potato foods as affected by cooking methods. *Int. J. Food Prop.* **2018**, *21*, 1875–1887. [CrossRef]
28. Grunenfelder, L.; Hiller, L.K.; Knowles, N.R. Color indices for the assessment of chlorophyll development and greening of fresh market potatoes. *Postharvest Biol. Technol.* **2006**, *40*, 73–81. [CrossRef]
29. Haase, N.U. Glycoalkaloid Concentration in Potato Tubers Related to Storage and Consumer Offering. *Potato Res.* **2010**, *53*, 297–307. [CrossRef]
30. Tilahun, S.; Park, D.S.; Taye, A.M.; Jeong, C.S. Effect of ripening conditions on the physicochemical and antioxidant properties of tomato (*Lycopersicon esculentum* Mill.). *Food Sci. Biotechnol.* **2017**, *26*. [CrossRef]
31. Tajner-Czopek, A.; Jarych-Szyszka, M.; Lisińska, G. Changes in glycoalkaloids content of potatoes destined for consumption. *Food Chem.* **2008**, *106*, 706–711. [CrossRef]
32. Percival, G.C.; Dixon, G.R. Glycoalkaloids. In *Handbook of Plant and Fungal Toxins*; Felix D'Mello, J.P., Ed.; CRC Press: Boca Raton, FL, USA, 1997; pp. 19–35.

33. Okamoto, H.; Ducreux, L.J.M.; Allwood, J.W.; Hedley, P.E.; Wright, A.; Gururajan, V.; Terry, M.J.; Taylor, M.A. Light Regulation of Chlorophyll and Glycoalkaloid Biosynthesis During Tuber Greening of Potato S. tuberosum. *Front. Plant Sci.* **2020**, *11*, 1–19. [CrossRef]
34. Arias, R.; Lee, T.C.; Logendra, L.; Janes, H. Correlation of lycopene measured by HPLC with the L*, a*, b* color readings of a hydroponic tomato and the relationship of maturity with color and lycopene content. *J. Agric. Food Chem.* **2000**, *48*, 1697–1702. [CrossRef]
35. Helyes, L.; Pék, Z.; Lugasi, A. Tomato fruit quality and content depend on stage of maturity. *HortScience* **2006**, *41*, 1400–1401. [CrossRef]
36. Spoladore, D.S.; Zullo, M.A.T.; Teixeira, J.P.F.; Coelho, S.M.B.M.; de Silva Miranda Filho, H. Synthesis of chlorophylls and glycoalkaloids in mature potato tubers stored under natural light. *Bragantia* **1985**, *44*, 197–208. [CrossRef]

MDPI
St. Alban-Anlage 66
4052 Basel
Switzerland
Tel. +41 61 683 77 34
Fax +41 61 302 89 18
www.mdpi.com

Horticulturae Editorial Office
E-mail: horticulturae@mdpi.com
www.mdpi.com/journal/horticulturae

www.ingramcontent.com/pod-product-compliance
Lightning Source LLC
LaVergne TN
LVHW070617170726
843515LV00004B/110

* 9 7 8 3 0 3 6 5 3 7 8 5 6 *